STOCHASTIC PROCESSES AND THEIR APPLICATIONS

STOCHASTIC PROCESSES AND THEIR APPLICATIONS

Ivan Stanimirović

ARCLER
PRESS

www.arclerpress.com

Stochastic Processes and their Applications

Ivan Stanimirović

Arcler Press

2010 Winston Park Drive,

2nd Floor

Oakville, ON L6H 5R7

Canada

www.arclerpress.com

Tel: 001-289-291-7705
001-905-616-2116
Fax: 001-289-291-7601
Email: orders@arclereducation.com

ISBN: 978-1-77361-378-9 (Hardcover)

ABOUT THE AUTHOR

Ivan Stanimirović gained his PhD from University of Niš, Serbia in 2013. His work spans from multi-objective optimization methods to applications of generalized matrix inverses in areas such as image processing and computer graphics and visualisations. He is currently working as an Assistant professor at Faculty of Sciences and Mathematics at University of Niš on computing generalized matrix inverses and its applications.

TABLE OF CONTENTS

LIST OF FIGURES

LIST OF TABLES

PREFACE

Stochastic processes are successions of events governed by laws of probabilıty. Many applications of stochastic processes appear in physics, engineering, biology, medicine and other disciplines as well as in other branches of mathematics. The purpose of these notes is to present a set of notions and useful to work with stochastic processes results. The presentation is not used measure theory but adopts a "constructive" view.

As it is known, Markov chains are short memory processes in the sense that only "remember" the last state visited to decide what next. Formally, the {Xn} process with state space E is a Markov chain if

$$P(X_{n+1} = y \mid X_n = x_n, \dots, X_0 = x_0) = P(X_{n+1} = y \mid X_n = x_n)$$

where an $x_0, \dots, x_n \in E$. In processes with long "memory" the value taking the process in each step depends on the entire past. This type of process was introduced by Onicescu and Mihoc [50] under the name chains with full hookups in 1935. The first chapter of this book is an introduction to this type of process where there is a possible construction of them and some results.

In the third chapter Markov chains are introduced continuous time and important results of these processes are demonstrated. For the development of this chapter a previous study of Poisson processes, which is made in the second chapter, is needed. In this chapter not only Poisson processes are defined and their basic properties are studied, but also the concept of process two-dimensional Poisson is introduced. This type of process is used in the development of the remaining sections to build one-dimensional homogeneous processes. The construction of Poisson processes performed in the second chapter is a particular case of the construction given by Neveu (1977) [45]. The construction process as smaller dimension greater projection dimension process was introduced by Kurtz (1989) [40]. The biggest difference between Markov chains and discrete time Markov chains continuous time is that the first jump from one state to another in integers 1, 2,... continuous time and do it at random times.

The fourth chapter is to show some of the applications that exist for the processes described in previous chapters. Show examples are related-two biology but those processes have many applications in areas such as other economics, sociology, physical, etc.

Finally they have been appended to the end of this work two appendixes. The first de-scribe certain type of process used in the development of these notes and some properties. In the second appendix of the generating functions defined and are some of the properties of these that are used in Chapter 4.

Service centers with networked care systems live day-to-day problems of congestion, inefficiency and lack of planning in the proper design of their model of care. Both conception and redesign involves a study based on performance measures that are affected by many variables constantly in and out of the system.

Each of these variables impact somehow the way the system behaves and how it is able to meet customer needs efficiently. Especially the human resource is the key factor in good design because it has a greater impact on measures of assessment and therefore represent the best starting point for implementing process improvement, study and analysis is only possible through statistical tests demonstrating their effect on the indicators around interactions with other factors affecting a queuing model.

This study is responsible for conducting this analysis combining different types of factors such as attendance rates arrival, probability distributions, length of a serial network, number of servers and system capacity to model queuing problems in different applications and significantly different results.

In addition from the obtained analysis, it is possible to construct an optimization algorithm that involves knowledge of the factors and their influence on indicators, so it's wise in choosing the best set of solutions to the problem, i.e., the number of servers and the capacity to meet customer needs. Techniques multi-objective optimization is very useful when minimize competing goals where there are no parameters that optimize both objectives simultaneously.

The inclusion of techniques known in operations research and optimization algorithm conceived from game theory, developed processes local search and evaluated by economic principles of equality, provides an analysis tool untraditional and innovative with different conclusions about the optimal state of a system.

CHAPTER 1

STOCHASTIC PROCESSES, MARKOV CHAINS AND CONTINUOUS-TIME APPLICATIONS

CONTENTS

In this chapter, we will start by introducing the notion of specification and in the second section describes how to build a stochastic process that is "compatible" with a specification. In the last section we will see some results for these processes. The concept of specification is used in statistical mechanics.

1.1. INTRODUCTION

Currently the partial differential equations have many applications in applied mathematics and physics. In the latter, these equations are used to describe linear phenomena.

In addition, for example, the application of new mathematical theories as stochastic calculus enabled the emergence of new both ordinary differential equations as partial derivatives, called stochastic differential equation [11]. In this sense, the occurrence of partial differential equations in stochastic calculus, has led to the application of some methods of resolution images that appear in mathematical physics and in some areas of applied mathematics. Particularly in financial mathematics, by generating financial mathematical models, such as the valuation of options.

On the other hand, they have generated many solutions of the main equations that appear in financial mathematics; these solutions were obtained by mathematical techniques used in Burke (1976) physical; but the problem lies in the fact that not all solutions obtained by these techniques are in correspondence with some applications of financial mathematics. Therefore, we have to make certain considerations in applying such solutions to this case. Such as partial differential equation Black-Scholes. It has its origins in financial mathematical models; particularly in the solution of option pricing.

But this equation has been solved by techniques of mathematical physics such as quantum mechanics and classical mechanics. In the case of quantum mechanics, this equation has a solution analogous to Feynman-Kac formula (Francis Comets et al., 2002); and in the case of classical mechanics are used techniques of separation of variables and theory of Lie groups through the introduction of symmetries and conservation laws. All these solutions obtained from (1) are analytical; which have the following problems, among others:

1. Some solutions do not meet initial value problems and boundary appearing in the financial markets (Burke, 1976).
2. Some solutions cannot be calculated using computational

techniques; preventing its application in solving some mathematical problems encountered in these markets.

Thus the aim of this chapter is to use this calculation for the partial differential equation (1), which is used to calculate the value of options in financial mathematics. Furthermore, since this equation can be reduced to a process of stochastic diffusion, by suitable transformation, this allows us to model valuation of options using the finite difference method (Kendall, 1952; Lotka, 1925), to thereby be able to compare the said calculating with the values obtained by the Black-Scholes formula, which is the analytical solution of the problem of the boundary value associated with the equation (1):

$$C(T,x) = \acute{\max}(x - K, 0),$$

$C(t, x) = x - Ke^{-r(T-t)}$ when $x \to \infty$, $C(t, 0) = 0\ \forall t$,

Additionally, compare the new solutions of the partial differential equation (1), obtained by symmetries with the Black-Scholes formula.

1.1.1. Random Variables

In this section we introduce some basic notions of probability theory necessary for the definition of Stochastic and back-mind for the definition of stochastic integrals processes.

In practice, two types of phenomena or experiments occur: The first we can name are deterministic phenomena which are determined-das actions which can predict what will happen. Unlike these, they are random phenomena which are actions that do not know what will happen to be performed, as can be obtained any of several alternatives, even under the same conditions in which realize these actions. Using the theory of probability, which we are interested in are increasing, it could make the study of random phenomena. Quickly we give some definitions that will allow us to explore the development of this theory.

Definition 1.1. Given a random phenomenon, the set of all possible outcomes is called the sample space and denote by Ω.

Definition 1.2. It is a random phenomenon with sample space Ω, a random variable X is a function whose domain is Ω and counter-domain is R:

$X: \Omega \to R.$

In addition to these two definitions, we call event to any subset of the sample space and define the F class as the collection of all events Ω, which will be described more accurately.

With these definitions, the probability measure can be seen as follows. Let F class events Ω, then for each event∈ F is assigned a number P (A) ∈[0, 1]. This number is the expected fraction of occurrence of the event A, along a series of experiments where A and Ac are observed.

Some elementary properties of probability measures are summed below.

i) It is satisfied that:

$P(\Omega) = 1$ and $P(\emptyset) = 0$.

ii) For events A, B ∈ F,

$P(A \cup B) = P(A) + P(B) - P(A \cap B)$.

iii) If A and B are disjoint, i.e., $A \cap B = \emptyset$, so

$P(A \cup B) = P(A) + P(B)$.

Note also that Property (iii) it follows that P (Ac) = 1 – P (A). The relationship between random variables and probability may be characterized by some numeric quantities. Now consider some of these.

Definition 1.3. The collection of odds

$F_X(X) = P(X \leq x) = P(\{\omega: X(\omega) \leq x\}), x \in R, \omega \in \Omega$

is the distribution function cumulative probability F_X X.

Example 1.1. Using the definition [1.3] we can determine probabilities in terms of the distribution function cumulative probability, such as:

1. $P(\{\omega: a < X(\omega) \leq b\}) = F_X(B) - F_X(A)$ with $a < b$.

2. $P(X = x) = F_X(X) - F\ \lim_X(X - h)$.

$h \rightarrow 0$

With these probabilities can approximate the probability of the event $\{\omega: X(\omega) \in B\}$ for any complicated set $B \subset R$.

Definition 1.4. The collection of odds

$$P_X(B) = P(X \in B) = P(\{\omega: X(\omega) \in B\}),$$

It is the distribution of X to $B \subset R$ adequate.

Note 1.1. Denoted by B to appropriate subsets of R^n. These subassemblies are Borel sets, which are obtained by a countable number of operations ∩,∪ or ^{c} acting on the intervals R^n.

The distribution P_X and the probability distribution function F accumulated notions are equivalent in the sense that both can be used to calculate the probability of any event $\{X \in B\}$.

A probability distribution function is continuous or discrete (it jumps). Consider first the special case when the distribution function F_X It is a discrete function.

$$pk = P(X = xk).$$

The distribution function of cumulative probability (1.1) and the corresponding distribution are said to be discrete. A random variable with distribution function cumulative probability (1.1) is a discrete random variable.

In contrast distributions and discrete random variables, the distribution function a continuous random variable has no jumps, therefore $P(x) = 0$for all x, or equivalently,

$$lim\, F_X(x + h) = F_X(x), \quad \forall x,$$

$$lim\, F_X(x + h) = F_X(x), \quad \forall x, \qquad (1.2)$$

i.e., such random variable assumes any value with probability 0. A random variable gets its name property continuity (1.2) of the probability distribution function F accumulated.

Example 1.2. (The normal distribution.)

An important continuous distribution is normal or Gaussian distribution N (μ, σ^2) With parameters $\mu \in R$, $\sigma > 0$. This function has the following probability density X~ N (0, 1), we say that the random variable X has a standard normal distribution, and φ denote with its probability density function f(x) and its distribution function Φ cumulative probability F_X.

On the other hand, some interesting characteristics of a random variable are hoping the variance and moments

Below we define these characteristics for both discrete variables as continuous variables.

We can consider the expectation μ_X as the center of gravity of the random variable X, that is, the random values X (ω), with $\omega \in \Omega$, which they are concentrated around the random value μ_X. The extension or dispersion of random values X (ω) around the expectation $\mu_{X/}$ It is described by the variance σ_{X2} and standard deviation σ_X.

Definition 1.9. We say that $X = (X_1..., X_n)$ is a vector of dimension n if their components $X_1..., X_n$ random variables are one-dimensional.

Analogously to the random variables of dimension we can enter the function of probability distribution, Expectation, moments, the matrix of variance-covariance of a random vector, in order to describe the distribution and structure dependent.

As before, it is considered a collection of subsets F

Ω and a probability measure in this, that is, assigning a number P (A) $\in [0, 1]$ is defined for each $A \in F$.

Definition 1.10. The collection of

$$F_X(\boldsymbol{x}) = P(X1 \leq x1, \ldots, Xn \leq xn)$$
$$= P(\{\omega : X1(\omega) \leq x1, \ldots, Xn(\omega) \leq xn\}),$$
$$\boldsymbol{x} = (x1, \ldots, xn) \in R^n,$$

function is the joint probability distribution F_X (X).

This provides us with the probability of the event that X assumes values in the rectangle

$$(\boldsymbol{a}, \boldsymbol{b}] = \{\boldsymbol{x} : ai < xi \leq bi, i = 1, .., n\}$$

As in the case of random variables of dimension, these probabilities approximate P ($X \in$ B) for very general sets B.

Definition 1.11. The collection of

$$P_X(B) = P(\boldsymbol{X} \in B) = P(\{\omega : \boldsymbol{X}(\omega) \in B\}) B \subset R^n,$$

constitute the distribution of X.

In a mathematical sense, the distribution and the distribution function of a joint probability random vector X are equivalent notions. Am-bas, F_X and p_X, They can be used to calculate the probability of any event $\{X \in B\}$.

Analogously random variables can introduce random vectors and discrete and continuous distributions of probability. For our purposes, the continuous random vectors with density will be relevance's, and therefore restrict our attention to these. If a vector X has probability density function , All-speakers with X_i, Vectors pairs), Triplets , etc., they have a probability density function. These are called functions density marginal probability.

In addition, the expectation of a random vector is similar as the mean value of a random variable function. X (ω) values are concentrated around the mean value. Thus we consider the following definitions.

Definition 1.13. X is a random vector.

i) Expectation or mean of X is given by:

$$\mu_X = E(\boldsymbol{X}) = (E(X1),\ldots,E(Xn)).$$

ii) The variance-covariance matrix of X is defined by:

$$\Sigma_X = (cov(Xi,Xj)\ ;\ i,j\ =\ 1,\ldots,n),$$

It is convenient to standardize covariance of the random variables by division by their corresponding standard deviations. As a result of this standardization, the correlation of two random variables is always between –1 and 1.

Example 1.3. (Gaussian random vector.)

A normal or Gaussian random vector has probabilistic distribution function normal or Gaussian accumulated. The distribution function or Gaussian normal probability of dimension n is given by the probability density function with parameters $\mu \in R^n$ and Σ, a positive definite symmetric matrix of size n × n, where Σ^{-1} is its inverse and its determinant $det\Sigma$. The parameter μ is the expectation μ_X X and Σ is the covariance matrix Σ-variance. Thus the probability density function of Gaussian vector, and hence its probability distribution function, are completely determined through expectation and variance-covariance matrix.

1.1.2. Independence and Dependence

Intuitively, if we have two experiments, independence means that the first experiment does not influence the outcome of the second and vice versa. In this section we will give some essential definitions and properties of independent events and independent random variables.

Definition 1.14. Two events A_1 and A_2 are independent if

$$P\,(A1 \cap A2)\ =\ P\,(A1)P\,(A2).$$

Definition 1.15. Two random variables X_1 and X_2. They are independent if

$$P\,(X1 \in B1, X2 \in B2) = P\,(X1 \in B1)\,P\,(X2 \in B2)\,, B1, B2 \subset R.$$

This means that the events $\{X_1 \in B_1\}$ and $\{X_2 \in B_2\}$ are independent.

Alternatively, it can be defined independently by distribution functions joint probability and probability density. Random variables X_1 and X_2 are independent if and only if

$F_{X1,X2}(x_1, x_2) = F_{X1}(x_1)\, F_{X2}(x_2), x_1, x_2 \in R.$

Suppose that (X_1, X_2) is a function of the joint probability density $f_{X1, X2}$ with marginal density functions probability $f_{X1}\, f_{X2}$. Then it can prove that the random variables X_1 and X_2 are independent if and only if

$f_{X1,X2}(x_1, x_2) = f_{X1}(x_1) f_{X2}(x_2), x_1, x_2 \in R.$

The definition of independence can be extended to a finite number of events and arbitrary random vectors. The independence of the components of a random vector implies independence of each pair of components, but the reverse is generally not true.

Definition 1.16. Events $A_1,..., A_n$ are independent if for any choice of indices $1 \le i_1 <... <$ Ny integer $1 \le k \le n$, is satisfied

$$P(A_{i_1} \cap \ldots \cap A_{i_k}) = P(A_{i_1}) \cdot\cdot\cdot P(A_{i_k}).$$

Definition 1.17. Random variables $X_1..., X_n$ are independent if for any choice of indices $1 \le i_1 <... < i_k \le n$, $1 \le k \le$ integers n subsets $B_1..., B_n$ R, it is satisfied

$$P(X_{i_1} \in B_{i_1}, \ldots, X_{i_k} \in B_{i_k}) = P(X_{i_1} \in B_{i_1}) \cdot\cdot\cdot P(X_{i_k} \in B_{i_k}).$$

This means that the events $\{X_1 \in B_1\},..., \{X_n \in B_n\}$ They are independent.

Random variables $X_1..., X_n$ They are independent if and only if its distribution function joint probability can be written as follows:

$F_{X^{\prime}...X}(x_1,..., x_n) = F_X(x_1) \cdot\cdot\cdot F_X(x_n), (x_1,..., x_n) \in R^n.$

If the random vector $X = (X_1..., X_n)$ Has probabilistic density function f Then we can prove that $X_1..., X_n$ are independent if and only if

$F_{X1..., Xn}(x_1,..., x_n) = F_{X1}(x_1) \cdot\cdot\cdot F_{Xn}(x_n), (X_1,..., x_n) \in R^n.$

An important consequence of random variables is the following property:

Property 1.1. If $X_1..., X_n$ They are independent, then for any real functions $g_1,..., g_n$, is fulfilled

$$E[g1(X1) \cdots gn(Xn)] = E[g1(X1)] \cdots E[gn(Xn)]$$

as long as deemed expectations are well defined.

In particular, we can conclude that the independent random variables X_1 and X_2 are uncorrelated, that is, $(X_1, X_2) = (X_1, X_2) = 0$. The converse is not generally true.

In the subsequent, often we treat with infinite collections (X_t, t ∈ T) of random variables X_t, That is, T is an infinite set of indices. In this arrangement, even independence will be introduced.

Definition. The collection of random variables (X_t, t ∈ T) is independent if for every choice other than indices $t_1,..., t_n \in T \geq 1$ and n random variables $X_{t1}..., X_{tn}$ are independent. This collection is independent and identically distributed (iid) if this is independent and random variables X have the same distribution.

Definition 1.19. A stochastic process X is a collection of random variables $(Xt, t \in T) = (Xt\ (\Omega), t \in T, \omega \in \Omega)\ defined$ in any space Ω.

For our purposes, T is often a range, for example T = [a, b], [a, b) or [a, ∞) for a <b. Then we call X a process of continuous time in contrast to a discrete time process. In the latter case, T is a finite or infinite countable set. For obvious reasons, the index t of a random variable X_t It is often referred to time, and follow this convention.

Definition 1.20. A stochastic process X is a function of two variables.

For a fixed time t, this is a random variable:

$$Xt = Xt\ (\Omega), \omega \in \Omega$$

For a random fixed output ω ∈ Ω, this is a function of time:

$$Xt = Xt\ (\Omega), t \in T.$$

This function is called one embodiment, a path or a path sample process X

We need the concepts of random variable and stochastic process X (X_t, t ∈T) are not very different. Both have paths, but the path X (ω) with ω∈ Ω of a random variable is a number, while the X Path (Ω), t ∈T, a stochastic process is a function in T. Thus correct it would be to understand a stochastic process as a random element taking functions as values. Moreover, we can interpret a random variable and a random vector as a special stochastic process with a finite set of indexes T.

Analogously random variables and random vectors introduced nonrandom characteristics of a stochastic process such as its distribution, Expectation, etc. desired and describe its dependence structure. This is a much more complicated than the description of a random vector task. Indeed, a nontrivial stochastic process X = (X_t, t ∈T) with an infinite set of indices T is an object of infinite dimension; this can be treated as an infinite collection of random variables X_t, t ∈T. Since the values of X are functions of T, the distribution of X may be defined on a subset of the feature space, i.e.,

$$P(X \in A), A \in F, \qquad (1.4)$$

where F is a collection of appropriate subsets of this space functions. The key observation is that a stochastic process can be interpreted as a collection of random vectors.

Definition 1.21. The finite-dimensional distributions (fidis) of a stochastic process X distributions are finite dimensional vector

$$(Xt1 \ldots, Xtn), \qquad T1\ldots, tn \in T,$$

for all possible choices of times $t_1, \ldots, t_n \in T$ and each $n \geq 1$.

Fidis can conceptualize much easier than the complicated distribution (1.4) of a stochastic process. It can be shown that fidis-end of the distribution of X. In this regard, we refer to the collection of fidis as the distribution of stochastic processes.

Stochastic processes can be classified according to different criteria. One is by the type of its Fidis.

Example 1.4. (Gaussian process.)

A stochastic process is called Gaussian if all its fidis are multivariate Gaussian with probability density function (1.3). Example 1.3 has the μ and Σ a Gaussian vector parameters are expectation and variance-covariance matrix, respectively. Therefore, the distribution of a Gaussian stochastic process is determined only by the collection of expectations and variance-covariance matrices of their fidis.

A simple Gaussian process on T = [0, 1] consists of variable N (0, 1). In this case the fidis are characterized by distribution functions

$$P(Xt_1 \leq x1, \ldots, Xt_n \leq xn) = P(Xt_1 \leq x1) \cdots P(Xt_n \leq xn)$$
$$= \Phi(x1) \cdots \Phi(xn),$$

where $0 \leq t1 \leq \cdots \leq tn \leq 1, (x1, \ldots, xn) \in R^n$

For a random vector X = (X_1..., X_n) Define expectation μ_X = (E (X_1),..., E (X_n)) And covariance matrix Σ- = ((X_i, X_j), I, j = 1,..., n). A stochastic process X = (X_t, t ∈ T) can be considered as the collection of all random vectors (X_{t1}..., X_{tn}) to T_1..., t_n ∈T n ≥ 1. For each can determine expectation and variance-covariance matrix. Alternatively we consider these quantities as functions of t∈ T:

Definition 1.22. Let $X = (X_t, t \in T)$ a stochastic process.

i) The function of expectation of X is given by:

$$\mu X(t) = \mu X_t = E(Xt), t \in T.$$

ii) Covariance function X is defined as:

$$cX(t,s) = cov(Xt,Xs) = E[(Xt - \mu X(t))(Xs - \mu X(s))], \ t, s \in T.$$

iii) The variance function of X is defined by:

$$\sigma_X^2(t) = cX(t,t) = var(X_t), t \in T.$$

As for a random vector, the function of expectation μX (t) is a deterministic quantity around which the trajectories of X are concentrated. CX covariance function (t, s) is a measure of dependence in the process X. The variance function σX2 (t) can be considered as a measure of dispersion of a path around μX X (t).

Another form of classification of stochastic processes is the imposition of a special structure of dependency.

Definition 1.23. The process $X = (X_t, t \in T)$, $T \subset R$, is strictly dependent if are invariant under changes of the indices t

$$(X_{t_1}, \ldots, X_{t_n}) \stackrel{d}{=} (X_{t_1+h}, \ldots, X_{t_n+h}),$$

Note 1.2. Here = means the identity of the distributions. For random vectors (1.5), this means that their distribution functions are identical.

If a process of real life by a stationary stochastic process described, then it would be considered that characteristic properties of this process do not change as time passes. The dependence structure described by fidis or covariance function is invariant by changes over time, which is a restriction on mind-strongest on the underlying process, however, this is a standard supposing in many fields related with probability, such as statistical and time series analysis.

The stationary property can also be imposed on the process increases. The process itself is then not necessarily stationary.

Definition 1.24. Let $X = (X_t, t \in T)$ a stochastic process and $T \subset R$ interval. X is said to have stationary increments if

$$X - X \stackrel{d}{=} X_{t+h} - X_{s+h} \text{ for all } s, t \in T \text{ with } t + h, s + h \in T.$$

X is said to have independent increments if for each choice of $t_i \in T$ with $t_1 < \cdots < T_n$

Xt2 - Xt1 ,...,Xtn - Xtn-1,

are independent random variables.

1.1.3. The Conditional Expectation Under Discrete Conditions

We know that the conditional probability of A given B happens is:

$$P(A \mid B) = \frac{P(A \cap B)}{P(B)}, \quad P = 0.$$

Clearly, P (A | B) = P (A) if and only if A and B are independent.

The probability P (A | B) can be interpreted as follows. Assume that the B event occurs. This is additional information which substantially changes the underlying probability measure. In particular, the probability is assigned to B 0^c (We know that B^c not occur) and 1 to B. The B event becomes our new probability space Ω^0. All events of interest are now subsets of Ω^0 : $A \cap B \subset \Omega^0$. In order to obtain a new probability measure on Ω^0it has to normalize old probabilities P (A ∩B) P (B). In short, the occurrence of B causes our original space is reduced to Ω Ω^0 and the original probabilities P (A) have to be replaced with P (A | B).

Since P (B)> 0, we can define the distribution function conditional probabilisticlity of a random variable X given B, such as:

$$F_X(x \mid B) = \frac{P(X \le x, B)}{P(B)}, \quad x \in \mathbb{R},$$

and even the conditional expectation of X given B denotes the indicator function of event B. In order to discuss the equation (1.6), assume for the moment that Ω = R. If X is a discrete random variable with values x_1, x_2,

Example 1.5. (The conditional expectation of a random variable uniform)

Consider the random variable X (ω) = ω in the space [0, 1] equipped with the probability measure P such that

P ((a, b]) = b – a, (A, b] ⊂ (0, 1].

Clearly, X has a uniform (0, 1] distribution and its function. Given that f_X (X) = 1 on (0, 1] and also that P (A_i) = 1/n, the conditional expectations E (X | Ai). The value E (X | Ai) is updated in the new space Ai expectation, given the information Ai occurred.

Now consider a discrete random variable Y on Ω taking different values and in sets A_i, that is to say,

$A_i = \{\Omega: Y(\omega) = y_i\}$ I = 1, 2,...

Clearly, (A_i) is a partition of Ω,

It is further assumed for convenience that $P(A_i) > 0$ for all i.

Definition 1.25. For a random variable X on Ω with E | X | <∞, the conditional expectation of X is defined as Y, a discrete random variable, as

$E(X|Y)(\omega) = E(X|A_i) = E(X|Y = y_i), \quad \forall\, \omega \in A_i,\ i = 1, 2, \ldots.$

If we know that A_i happens, we can stick to the ω's in A_i. For those 'S, E (X | Y) (ω) coincides with the conditional expectation E (X | A_i).

Example 1.6. (Continuation of Example 1.5.)

Interpret the E (X | M) 's in (1.7) the values of a discrete random variable E (X | Y), where Y is constant over the sets Ai = ((i – 1)/n, i/n] for an illustration of this random variable in this E(X | Y) is nothing but a version of the random variable approximate original X, given the information that any of the RAs occur.

In what follows some basic properties of the variable E (Y|X) is given.

Theorem 1.1. The conditional expectation is linear, i.e., for variables X_1, X_2 and constant c_1, c_2 we have to:

$E([c_1X_1 + c_2X_2]|Y) = c_1E(X_1|Y) + c_2E(X_2|Y).$

Theorem 1.2. Expectations of X and E (X | Y) are the same, i.e., E (X) = E [E (X | Y)].

Theorem 1.3. If X and Y are independent, then E (X | Y) = E (X).

Proof. Definition [1.15] follows that the independence of X and Y, involves

$$P(X \in A,\ Y = y_i) = P(X \in A)P(Y = y_i) = P(X \in A)P(A_i), \qquad (1.10)$$

considering the random variable I_{TOi} and noting that

$\{\omega : I_Ai(\omega) = 1\} = A_i = \{\omega : y(\omega) = y_i\},$

can rewrite (1.10) as follows

$P(X \in A,\ I_{Ai} = 1) = P(X \in A)P(I_{Ai} = 1).$

The analogous relationship, where $\{I_{Ai} = 1\}$ is replaced with $\{I_{Ai} = 0\}$ is satisfied even.

Therefore the random variables X and I_{TOi} They are independent and ω ∈ TO_i, We must:

$$E(X|Y)(\omega) = E(X|A) = E(X),$$

where we use

$$E(IA) = 0 \cdot P(A^C) + 1 \cdot P(Ai) = P(Ai).$$

Summing up the most important points of this section:

The conditional expectation E (X | Y) of X given a discrete random variable

It is a discrete random variable.

This coincides with the classical conditional expectation E (X | Y = y_i) In the sets A_i = {Ω: Y (ω) = y_i}.

In this sense this is an approximate version of X.

In particular, if Y is equal to a constant, then E (X | Y) = E (X), if Y assumes two different values, so E (X | Y), etc.

The conditional expectation E (X | Y) is not a function of X is simply a function of Y. The random variable X only determines the type of function. Indeed, we can write

1.1.4. σ-Algebras

(X | Y) the conditional expectation E was introduced a random variable X under the discrete condition (i.e., the discrete random variable) Y. Remembering (1.8), the values of Y do not really matter to the definition of E (X | Y); but it was crucial that we assume different values within ω-sets A_i. Thus, the conditional expectation E (X | Y) can actually be understood as a random variable constructed from a collection σ (Y) of subsets of Ω. Therefore it could, in a simplified manner, write E (X | σ (Y)).

Obviously, the collection σ (Y) provides information about the structure of the variable Y (ω) as a function of ω∈ Ω.

In what follows, we want to make more precise meaning the collection σ (Y) of subsets of Ω. The call σ-field or σ-algebra. Its definition is as follows:

Definition 1.26. A σ-algebra F (in Ω) is a collection of subsets of satisfying the following conditions:

i) It is not empty;∅ ∈ F and Ω ∈ F.

ii) If A ∈ F, then A^c ∈ F.

Example 1.7. (Elementary Σ-algebras.)

The following collection of subsets of Ω Examples of σ-algebras:

1. $F_1 = \{\emptyset, \Omega\}$.
2. $F_2 = \{\emptyset, \Omega, A, A^c\}$ For some Af=∅ and Af= Ω.
3. $F_3 = P(\Omega) = \{A: A \subset \Omega\}$.

F1 is the smallest σ-algebra in Ω and F3, The power set of Ω, is the largest, because it contains all possible subsets of Ω.

Now, suppose that C is a collection of subsets of Ω, but not necessary one σ-algebra. Adding more sets of C, always may have a σ-algebra, for example, the power set P (Ω). However, there are reasons mathematics show that P (Ω) is generally too large. But we can even try:

Observation 1.1. C for a collection of subsets of Ω, there is a σ-algebra σ (C) smaller in Ω containing a C.

We call σ (C), the σ-algebra generated by C.

Example 1.8. (Σ-algebras generated from collections of sub-sets Ω.)

Recalling the σ-algebras of Example 1.7 we have $F_i \; \Sigma = (C_i)$

1. If as $C_1 = \{\emptyset\}$, Is added to Ω C_1.
2. If we are given $C_2 = \{A\}$, A are added, ∅ and Ω to C_2.
3. If we are given $C_3 = F_3$, It is met by the definition of F_3.

In general, it is difficult, if not impossible, to give a description of the constructive elements σ (C). The σ-algebra σ (Y) generated by a discrete random variable Y is an exception as shown below.

Example 1.9. (The σ-algebra generated by a random discrete variable.)

Considering a discrete random variable Y with different values and definite subsets $A_i = \{\Omega: Y(\omega) = y_i\}$, which constitute a disjoint partition of Ω. choose

$C = \{A_1, A_2, \ldots\}$.

where I is any subset of N = {1, 2,...}, Including I =∅ (Giving A = ∅) And R = N (giving A = Ω). Since the sets (1.11) necessarily belong to σ (C), the smallest σ-algebra containing C, even we have σ (Y) = σ (C). Then we will call σ (Y), the σ-algebra generated by Y.

Note that σ (Y) contains all sets of the form

$A_{a,b} = \{Y \in (A, b] = \{\omega\}: \text{to} < Y(\omega) \leq b\}, \qquad -\infty < a < b < \infty.$

In fact, the set $I = \{i: a < \text{and}_i \leq b\}$ it is a subset of N,

Example 1.10. (Borel sets.)

We have $\Omega = R$ and

$C^{(1)} = \{(A, b]: -\infty < a < b < \infty\}$.

The σ-algebra B1 = σ (C (1)) contains many general subsets of R. This is called σ-Borel set, its elements are Borel sets. For example, it is a fact, but it is not easy to verify that B1 is a subset itself of the power set P (R). We can even enter σ-algebras of sets n-dimensional Borel B_n $\Sigma = (C^{(N)})$ Where $\Omega = R^n$ Y

$C^{(N)} = \{(A, b]: -\infty < a_i < b_i < \infty, i = 1,..., N\}$.

Sets (a, b] are called rectangles. Any set R is a Borel set. For example, balls, spheres, smooth curves, surfaces, etc., are Borel sets, they thus are also open sets and closed. In order to show that given a subset $C \subset Rn$ is a set of Borel, C is necessary to obtain by a countable number

Now recall the example 1.9, where the σ-algebra generate σ (Y) of sets $A_i = \{\Omega: Y(\omega) = y_i\}$ of a discrete random variable. If Y is a random variable taking continuous values, the σ-algebra generated from the sets of the form $\{\omega: Y(\omega) = y\}$, and element of R, from a mathematical standpoint rich enough. For example, the sets of the form $\{\omega: a < Y(\omega) \leq b\}$ none such σ-algebra, but it is desirable to have them in σ (Y).

It was in Example 1.9 that, for a discrete random variable Y, the with-together $\{\omega: a < Y(\omega) \leq b\}$ belong to σ (Y). Therefore we require a minimum assumption that these sets belong to σ (Y) when Y is any random variable. Since we are even interested in random vectors and immediately define the σ σ-algebras (Y) for the multi-varied case.

Definition 1.27. Let $Y = (Y_1,..., Y_n)$ A random vector of dimension n, i.e., $Y_1,..., Y_n$ are random variables. The σ-algebra σ (Y) is the smallest σ-algebra containing all sets of the form

$\{Y \in (A, b] = \{\omega\}: a_i < Y_i(\Omega) \leq b_i, I = 1,..., N\}$,

$-\infty < a_j < b_j < \infty, j = 1,..., N$.

We call σ (Y) the σ-algebra generated by the random vector Y.

An element σ (Y) gives an information whether $\omega \in \Omega$ is a random vector Y, which takes values in a rectangle (a, b] or more generally a Borel set.

Note 1.3. The σ-algebra σ (Y) generated by Y contains the essential information about the structure of the random value and as a function of $\omega \in \Omega$. This contains all sets of the form $\{\omega: Y(\omega) \in C\}$ for all Borel sets $C \subset R^n$.

As seen, it is difficult to imagine the σ-algebra σ (Y), and it's even harder for a stochastic process.

Definition 1.28. For a stochastic process Y = (Yt, t∈ T, ω ∈ Ω), the σ-algebra σ (Y) is the smallest σ-algebra containing all sets of the form

{Ω: the trajectory (Y_t(Ω), t ∈ T) belongs to C},

for all appropriate set of functions in C T. Then σ (Y) is called the σ-algebra generated by Y.

We have learned that the σ-algebra σ (Y) is a non-trivial object. In what follows, we use the following rule in order to be able to imagine σ (Y).

Definition 1.29. For a random variable, random or stochastic vector and Ω process, the σ-algebra generated and contains the essential information about the structure of Y as a function of ω∈Ω. This consists of all subsets {ω: Y (ω)∈ C} for suitable sets C.

Since Y generates a σ-algebra, even we say that Y contains information represented by σ (Y) or Y carries the information of σ (Y).

We conclude this section with a useful observation. Let f be a function acting on Y, and consider the set

{Ω: f (Y (ω)) ∈ C},

for suitable sets C. For suitable functions f, this set belongs to σ (Y), that is,

σ (f (Y)) ⊂ σ (Y).

This means that a function f acting on and does not provide new information about the structure of Y. We say that the information contained by f (Y) is contained in the information σ (Y).

1.1.5. General Conditional Expectation

(X | Y) the conditional expectation E was defined a random variable X, given a discrete random variable Y. This definition does explicit use of yi values of Y, but this depends on the subset Ai = {ω: Y (ω) = yi} of Ω. is showed in Example 1.9 that the collection subsets A_i generate the σ-algebra σ (Y). This is the starting point for the definition of the conditional expectation E (X | F) given a σ-algebra F in Ω. In applications always take F = σ (Y) random variables, vectors random or stochastic processes and appropriate. In section 1.3.2 it was made clear that the essential information about the structure and contained in the -algebra σ (Y) generated by Y. In this sense, we say that Y has the information σ (Y).

Definition 1.30. Observe Y, Y_1, Y_2 random variables, random vectors or processes stochastic in Ω and F one σ-algebra in Ω.
We say that

i) And the information is contained in F or Y does not contain more information than is contained in F if σ (Y)⊂ F, and

ii) Y_2 It contains more information than Y_1 if σ (Y_1) ⊂ σ (Y_2).

We are now ready to give a rigorous definition of conditional expectation E (X | Y) under a σ-algebra abstract F.

Definition 1.31. A random variable Z is called expectation conditional X given the σ-algebra F (we write Z = E (X | F)) if

i) Z does not contain more information than is contained in F:

σ (Z) ⊂ F.

ii) Z satisfies the relation

$$E(XIA) = E(ZIA) \, \forall A \in F \tag{1.12}$$

Property defined by the equation (1.12) shows that the variables X and E (X | F) are close to each other, not in the sense that they coincide for any $\omega \in \Omega\omega \in \Omega$, but approximations (expectations) of X and E (X | F) on appropriate sets are the same. This supports our conceptualization of the random variable E (X | F), which we obtained in section 1.3.1.

Note 1.4. The conditional expectation is an approximate version of the original random variable X.

The property defined in (1.12) leaves some leeway for the construction of the random variable Z = E (X | F). This means that there may exist versions Z 'of E (X | F) which may differ from Z on joint probability 0. For this reason, all relationships that envelop E (X | F) should actually be interpreted in felt safer, and even this uncertainty should indicate any relevant formula (see Appendix B). However, once we pointed this fact, we find convenient to remove this tag wherever.

Example 1.11. (The conditional expectation under discrete conditions) We want to show that the definition of E (X | Y) in section 1.3.1 can be viewed as a special case of the definition [1.31] E (X | F) where F = σ (Y).

In (1.9) we define E (X | Y) as a random variable Z such that

Z (ω) = E (X | A_i) For ω ∈ A_i.

Recalling that Z is only a function of Y, not X, therefore σ by (Z)⊂σ (Y). Moreover, for A given by (1.13), Z satisfies the relationship defined in (1.12) and it contains no more reporting Y (is a function of Y). Therefore this is indeed expectation conditional X given the σ-algebra σ (Y).

In the case of a discrete random variable and we have seen that E (X | Y) and E (X | σ (Y)) represent the same variable. This suggests the following definition:

Definition 1.32. Let Y be a random variable, a random vector or stochastic process Ω and σ (Y) having a σ-algebra generated by AND. The conditional expectation of a random variable X given Y is defined by

E (X | Y) = E (X | σ (Y)).

Example 1.12. (The conditional probability and conditional expectation.) The conditional probability and the conditional expectation are special cases of the general notion of conditional expectation defined in the definition [1.31]. Indeed, B is such that P (B) 6 = 0, P (B^c) 6 = 0 and define $F_B = \Sigma (\{B\})$. We have $F_B = \{\emptyset, \Omega, B, B^c\}$ Example 1.11 and considering we have

E (X | F_B) (Ω) = E (X | B) for ω∈ B.

This is the classic notion of conditional expectation. If we specify that I_A for some event A, we obtain for ω ∈ B, right side is the conditional probability of A given B.

1.1.6. Rules for Computation of Conditional Expectations

Theorem 1.4. If E | X | <∞, the conditional expectation E (X | F) exists and is unique.

The proof of this theorem is beyond the scope of the tools used, so only use it as a fact to know that the conditional expectation is well defined. In order to use later this important concept, given below are some rules for the calculation of conditional expectations.

Rule 1. The conditional expectation is linear. Random variables X_1, X_2 and constant c_1, c_2, We have

$E([c_1X_1, c_2X_2]|F) = c_1E(X_1|F) + c_2E(X_2|F).$

Rule 2. Expectations X and E (X | F) are the same:

$E(X) = E[E(X|F)].$

Rule 3. If X and σ-algebra F are independent, then E (X | F) = $E(X)$. In particular, if X and Y are independent, then E (X | Y) = E(X).

Rule 4. If the σ-algebra σ (X) generated by the random variable X, is contained in F, then

E (X | F) = X.

Particularly if X is a function of Y, σ (X)⊂ σ (Y), so E (X | Y) = X.

Rule 5. If the σ-algebra σ (X) generated by the random variable X, is contained in F, then for every random variable G

E (XG | F) = XE (G | F).

In particular, if X is a function of Y, σ (X)⊂ σ (Y), so E (XG | Y) = XE (G | Y).

Rule 6. If F and F^0 are two σ-algebras with F ⊂ F^0, so

E (X | F) = E (E (X | F^0) | F)),

and

E (X | F) = E (E (X | F) | F^0)).

Rule 7. If X is independently F and the information given by a random variable, random or stochastic process vector G is contained in F, then for any function h (x, y)

E (h (X, G) | F) = E (E_X [H (X, G)] | F),

where E_X [H (X, G)] means that we look G and expectation with respect to X.

1.2. PRELIMINARY RESULTS

Consider a finite space E.

$N^* = N - \{0\}\ y - N^* = \{-i : i \in N^*\}$

sets the positive and negative integers, respectively.

Definition 1.1.1. We say that a function

$P: E \times E^{-N^*} \to [0,1]$

is a specification

$\underline{w} = (w_{-1}, w_{-2}, \dots) \in E^{-N^*}$

1. $P(y; \underline{w}) \geq 0$ for each $y \in E$
2. $\sum_{y \in E} P(y; \underline{w}) = 1$

Proposition 1.1.2. Given $\underline{w} = (w_{-1}, w_{-2}, \dots) \in E^{-N^*}$ an arbitrary configuration of *"It happened," it is possible to construct a stochastic pro-*

cess $\{X\frac{w}{t}:t\in Z\}$ with the property *that for all* $n\geq 0$ and arbitrary values $x_0, x_1, \dots x_n\in E$ is true:

$$P\left(X\frac{w}{t}=x_t:t=0,\dots,n \mid X\frac{w}{i}=w_i:i\in -N^*\right)=\prod_{t=0}^{n} P(x_t;\underline{\widehat{w_t}}) \quad (1.1)$$

where $\underline{\widehat{w_t}}=(x_{t-1},\dots,x_0,\underline{w})=(x_{t-1},\dots x_o,w_{-1},w_{-2},\dots))$

Proof. First note that it is possible to construct a family of sets from the interval [0; 1]

$$\{B(y;\underline{w}):y\in E\}:\underline{w}\in E^{-N^*} \quad (1.2)$$

that satisfies the conditions:

1. $|B(y,\underline{w})|=P(y;\underline{w})$
2. $\cup_{y\in E} B(y;\underline{w})=[0,1]$

for all $\underline{w}\in E^{-N^*}$. For this reaches to "order" elements E and then with length interval

$P(y;\underline{w})$

Property (2) in the definition of specification it allows us to do this construction.

Once the family partition (1.2) constructed we define the function

$F:E^{-N^*}\times[0,1]\to E$

as

$$F(\underline{w};u):=\sum_{y\in E} y\,1_{\{u\in B(y;\underline{w})\}} \quad (1.3)$$

$$X\frac{w}{0}=F(\underline{w};U_0)\quad t\geq 1$$

$$X\frac{w}{t}=F(X\frac{w}{t-1},\dots,X\frac{w}{0},\underline{w};U_t) \quad (1.4)$$

$\{U_n:n\in Z\}\}$ is a family of valid uniformly distributed in the range [0, 1].

$\{X\frac{w}{t}:t\geq 0\}$

is to see the process verifies the condition (1.1). Let

$n\geq 0$

and

$x_0,\dots,x_n\in E.$

Therefore

$$P\left(X\frac{w}{t} = x_t : t = 0, \ldots, n \mid X\frac{w}{i} = w_i : i \in -N^*\right) =$$

$$P\left(F\left(X\frac{w}{t-1}, \ldots, X\frac{w}{0}, \underline{w}; U_t\right) = x_t : t = 0, \ldots, n \mid X\frac{w}{i} = w_i : i \in -N^*\right) =$$

$$P(\sum\nolimits_{y\in E} y\, 1_{\{U_t \in B(y; \frac{w}{t-1} \ldots X\frac{w}{0}, \underline{w})\}} = x_t : t = 0, \ldots, n \mid X\frac{w}{i} = w_i : i \in -N'$$

and from this we have

$$\prod_{t=0}^{n} P(\sum\nolimits_{y\in E} y\, 1_{\{U_t \in B(y; X\frac{w}{t-1}, \ldots, X\frac{w}{0}, \underline{w})\}} = x_t \mid X\frac{w}{i} = w_i : i \in -N^*) =$$

$$\prod_{t=0}^{n} P(\sum\nolimits_{y\in E} y\, 1_{\{U_t \in B\left(y; X\frac{w}{t-1}, \ldots, X\frac{w}{0}, X\frac{w}{-1}\right)\}} = x_t) = \mathsf{I}$$

$$\prod_{t=0}^{n} P(\sum\nolimits_{y\in E} y\, 1_{\{U_t \in B\left(y; X\frac{w}{t-1}, \ldots, X\frac{w}{0}, X\frac{w}{-1}\right)\}} = 1) = \mathsf{I}$$

$$\prod_{t=0}^{n} |B(x_t; x_{t-1}, \ldots, x_0, w_{-1}, \ldots)| =$$

$$\prod_{t=0}^{n} P(x_t; x_{t-1}, \ldots, x_0, w_{-1}, \ldots) =$$

$$\prod_{t=0}^{n} P(x_t; \underline{\widehat{w_t}})$$

Where $\underline{\widehat{w_t}} = (x_{t-1}, \ldots, x_0, w_{-1}, w_{-2}, \ldots)$

We introduce some notations below and see a particular way of constructing the partitions B (¢; ¢) for specifying P.

Given $n \in N, y \in E$ and $\underline{w} \in E^{-N^*}$ we define:

$$a_k(y; \underline{w}) := \inf\{P(y; w_{-1}, \ldots w_{-k}; z) : \underline{z} \in E^{-N^*}\} \quad (1.5)$$

where (w.,; :::; w,; z) = (w.,; :::; w,; z.,; z.,; :::). note that $(w_{-1}, \ldots w_{-k}, z) = (w_{-1}, \ldots, w_{-k}, z_{-1}, z_{-2}, \ldots)$

therefore does not depend on w. Since N is defined, observe

$$a_0(y; \underline{w}) := \inf\{P(y; \underline{z}) : z \in E^{-N^*}\}$$

$$k \in N: a_k := \inf_{\underline{w}\in E^{-N^*}} (\sum\nolimits_{y\in E} a_k(y; \underline{w})) \quad (1.6)$$

For $m \in N$

$$\beta_m := \prod_{k=0}^{m} a_k \tag{1.7}$$

$$\beta_m := \lim_{m\to\infty} \beta_m \tag{1.8}$$

Definition 1.1.3. A P specification is said to be full hookups if .

Definition 1.1.4. We say that a measure is compatible with the specification P if

$$v(\underline{X} \in E^Z : X_n = y \mid X_{n+j} = w_j : j \in -N^*) = P(y; \underline{w}) \tag{1.9}$$

for all $n \in Z$ $y \in E$, $\underline{w} = (w_{-1}, w_{-2}, \dots) \in E^{-N^*}$

1.2.1. Construction Process

In this section we will see a construction process set out in the proposal 1.1.2. Let us assume that the state space

$$E = \{1,2,\dots,q\}$$

for some positive integer q.

Each $\underline{w} \in E^{-N^*}$ define: $\{B_k(y;\underline{w}) : y \in E, k \in N$ partition the interval [0; 1] with the following properties:

1. And for $k \geq 0$, $\{B_k(y;\underline{w})$ is closed by left and right open with Lebesgue measure range $b_k(y;\underline{w})$
2. Intervals are arranged in increasing lexicographic order such that the left end of an interval coincides with the right end of the above:
3. $B_0(1;\underline{w}), B_0(2;\underline{w}), \dots, B_0(q;\underline{w}), B_1(1;\underline{w}), \dots, B_1(q;\underline{w}), \dots$

   without intersections. More precisely, calling $L(A) := \inf\{x : x \in A\}$, $R(A) := \sup\{x : x \in A\}$, $A \in R$ above construction requires:

1. $L\left(B_0(1,\underline{w})\right) = 0.$
2. $R(B_k(y;\underline{w})) = L(B_k(y+1;\underline{w}))$ $y = 1,\dots,q-1\backslash$
3. $R\left(B_k(q;\underline{w})\right) = L\left(B_{k+1}(1;\underline{w})\right).$

The above properties imply that:

$$B(y; \underline{w}) := \cup_{k\geq 0} B_k(y; \underline{w}) \qquad (1.10)$$

Proof. Due to the construction of the partition in the interval [0; 1]. It has to:

$$R(B_k(q; \underline{w})) = \sum_{y\in E} a_k(y; \underline{w})$$

$$R\left(B_k(q; \underline{w})\right) = \sum_{i=0}^{k}\sum_{j=1}^{q} |B_i(j; \underline{w})|$$

$$= \sum_{i=0}^{k}\sum_{j=0}^{q} b_i(j; \underline{w})$$

$$= \sum_{i=0}^{k}\sum_{j=0(}^{q} (a_i(j; \underline{w}) - (a_{i-1}(y; \underline{w})) + \sum_{j=1}^{q} a_0(y; \underline{w})$$

$$= \Sigma_{j=1}^{q}(a_k(j; \underline{w}) - a_0(j; \underline{w})) + \Sigma_{j=1}^{q} a_0\,(j; \underline{w})$$

$$= \sum_{j=1}^{q} a_k(y; \underline{w})$$

3. *Proof:* Using the result of (1) we have:

$$\lim_{k\to\infty} R(B_k(q; \underline{w})) = 1$$

$$\lim_{k\to\infty} R(B_k(q; \underline{w})) = \lim_{k\to\infty} a_k(y; \underline{w})$$

$$= \sum_{y\in E} \lim_{k\to\infty} a_k(y; \underline{w})$$

$$= \sum_{y\in E} P(y; \underline{w})$$

$$= 1$$

where the last equality is a corollary of the definition of P.

$$|B(y; \underline{w})| = P(y; \underline{w})$$

$$B(y; \underline{w})| = |\bigcup_{k\geq 0} B_k(y; \underline{w})|$$

$$= \sum_{k \geq 0} |B_k(y; \underline{w})|$$

$$= \sum_{k \geq 0} b_k(y; \underline{w})$$

$$= \sum_{k \geq 1} (a_k(y; \underline{w}) - a_{k-1}(y; \underline{w})) + a_0(y; \underline{w})$$

$$= \lim_{k \to \infty} a_k(y; \underline{w})$$

$$= P(y; \underline{w})$$

$$1 = \sum_{y \in E} |B(y; \underline{w})| \leq \sum_{y \in E} P(y; \underline{w}) = 1$$

$$B(y; \underline{w})| = P(y; \underline{w}) \tag{1.11}$$

4. $| \bigcup_{y \in E} B(y; \underline{w})| = 1$

$$| \bigcup_{y \in E} B(y; \underline{w})| = \sum_{y \in E} |B(y; \underline{w})|$$

$$= \sum_{y \in E} P(y; \underline{w})$$

$$= 1$$

Proof. It follows from the properties of the partition made that:

$$|B(y; \underline{w})| = |\bigcup_{k \geq 0} B_k(y; \underline{w})|$$

$$= \sum_{k \geq 0} |B_k(y; \underline{w})|$$

$$= \sum_{k \geq 0} b_k(y; \underline{w})$$

$$= \sum_{k \geq 0} \left(a_k(y; \underline{w}) - a_{k-1}(y; \underline{w})\right) + a_0(y; \underline{w})$$

On the other hand we know that:

$$\lim_{k \to \infty} a_k(y; \underline{w}) \leq P(y; \underline{w})$$

the intervals are disjoint and property (3)

$$1 = \sum_{y \in E} |B(y; \underline{w}) \leq \sum_{y \in E} P(y; \underline{w}) = 1$$

where the last equality is because P is a specification.

We define for $l \geq 0$:

$$B_t(\underline{w}) := \bigcup_{y \in E} B_t(y; \underline{w})$$

Observation 1.2.1. 1. Note that does not depend on w for all and
$y \in E$
since is an interval length

$$b_0(y; \underline{w}) = a_0(y; \underline{w})$$

which does not depend on w. This implies that $B_0(\underline{w})$ does not depend on w. *It follows from the definition of* a_k both $B_k(y; w)$ and $B_0(\underline{w})$ depend only on the first k w coordinates as they are length intervals

So far we have built a partition with all the necessary properties. This is how to construct the process sought. Let

$$\underline{w} \in E^{-N^*}$$

condition arbitrary edge. For this condition we define the function

$$F: E^{-N^*} \times [0,1] \to E$$

such that

$$b_k(y; \underline{w}) = a_k(y; \underline{w}) - a_{k-1}(y; \underline{w})$$

$$F(\underline{w}, u) := \Sigma_{y \in E}\, y\, 1_{\{u \in B(y; \underline{w}\}} \qquad (1.12)$$

Let $\underline{U} := \{U_i\}_{i \in Z}$ a succession of uniformly distributed in the interval [0; 1]. Finally we define the process as $X\frac{w}{0} := F(\underline{w}, U_0)$ and $t \geq 0$

$$X\frac{w}{0} := F(X\frac{w}{t-1}, \ldots, X\frac{w}{0}, \underline{w}, U_t) \qquad (1.13)$$

Lemma 1.2.2. The process defined by 1.13 distribution is 1.1, i.e., the distribution is consistent with the specification P and edge condition of w.

Proof. Proposition 1.1.2 just have to verify that the intervals have the correct length, which is proved by the property (3) of the partition,

$$|B(y; \underline{w})| = P(y; \underline{w})$$

Let's use some properties that we will be useful later. Note first that

$$[0, a_k] \subset \cup_{l=0}^{k} B_t(\underline{w}) \;\; \forall \underline{w} \in E^{-N^*} \qquad (1.14)$$

1. $L(\bigcup_{l=0}^{k} B_t(\underline{w}) \leq 0$

2. $a_k \leq R(\bigcup_{l=0}^{k} B_t(w))$

Now, on the one hand we have:
$(L(\bigcup_{l=0}^{k} B_t(\underline{w})) = L(B_0(\underline{w})) = L\left(B_0(1;\underline{w})\right) = 0$

On the other hand we have:

$$R(\bigcup_{l=0}^{k} B_t(\underline{w})) = R(\bigcup_{l=0}^{k}\bigcup_{y\in E} B_t(y;\underline{w}))$$

$$= R(B_k(q;\underline{w}))$$

$$= \sum_{y\in E} a_k(y;\underline{w})$$

where the last equality is due to property (1) of the partition. Now, using

$$a_k = \inf_{\underline{w}\in E^{-N^*}} (\sum_{y\in E} a_k(y;\underline{w}))$$

$$\leq \sum_{y\in E} a_k(y;\underline{w})$$

$$= R(\bigcup_{l=0}^{k} B_t(\underline{w}))$$

which ends the test. 1.14 Noting get if $u \leq a_k$ then to know the value $F(\underline{w};u)$ need only consider $w_{-1}, w_{-2}, \ldots, w_{-k}$ since $u \in [0, a_k] \subset \bigcup_{l=0}^{k} B_l(\underline{w}))$. More precisely 1.14 we can deduce that if $\underline{w}, \underline{v} \in E^Z$ are such that : $w_j = v_j\ \forall j \in [-k, -1]$

$$[0, a_k] \cap B_t(\underline{w}) = [0, a_k] \cap B_l(\underline{w}) \tag{1.15}$$

Thus we have:

$$u \leq a_k \;\; F(\underline{w}, u) = F(\underline{v}, u) \tag{1.16}$$

For all $\underline{w}, \underline{v} \in E^{-N^*}$, where $w_j = v_j\ \forall j \in [-k, -1], y \in [0,1]$

1.2.2. Results

We will see in this section some valid for construction work done in the previous sections results.

For each $\underline{w}, \underline{v} \in E^{-N*}$ is:

$$\tau^{\underline{w},\underline{v}} := \inf\{n \geq 0: X\frac{w}{k} = X\frac{v}{k}, \forall k \geq n\} \tag{1.17}$$

It is clear that in principle under which that time that result.

Proposition 1.3.1. If $\prod_{k\geq 0} a_k > 0$ then for all w and v inE^{-N^*}. It is true that $\sum_{n\geq 0} P(\tau^{\underline{w},\underline{v}} = n) = 1$ $\tau^{\underline{w},\underline{v}}$ can be infinity. However, there are certain conditions is almost certain finite. The following proposition

For all$\underline{w},\underline{v} \in E^{-N^*}$ (1.18)

$$\tau^{\underline{w},\underline{v}} := \inf\{n \geq 0 : X\frac{w}{k} = X\frac{v}{k}, \forall k \geq n\} \quad (1.19)$$

In other words, $\tau^{\underline{w},\underline{v}}$ is dominated stochastically for the fourth quarter return to the origin in the process "house of cards" with transitions

$$Q(k, k+1) = a_k \;;\; Q(k, 0) = 1 - a_k \quad (1.20)$$

and initial state $0, k \in N$ (see Appendix A). Now, by Lemma 1.3 condition of the hypothesis, $\prod_{k\geq 0} a_k \geq 0$ is equivalent to the transience of the process "house of cards."

This implies that the last return must be in finite time with probability 1, which proves the proposition.

For $-\infty < s < t \leq +\infty$ we define:

$$\tau[s,t] := \max\{m \leq s : U_k < a_{k-m}, \forall k \in [m,t]\} \quad (1.21)$$

that observe $\tau[s,t]$ may be ∞. We note $\tau[n] = \tau[n,n] := \max\{m \leq n : U_k < a_{k-m}, \forall k \in [m,n]\}$

Observation 1.3.2. Fixed $-\infty<s<+\infty$ that [s is fulfilled; $[s,t]$ is not increasing with t, because if $t \leq t`$ is true:

$$\{m \leq s : U_k < a_{k-m}, \forall k \in [m,+\infty]\} \subseteq \{m \leq s : U_k < a_{k-m}, \forall k \in [m,t`]\}$$
$$\subseteq \{m \leq s : U_k < a_{k-m}, \forall k \in [m,t]\}$$

Then taking the maximum of each set we have:

$$\tau[s,+\infty] \leq \tau[s,t`] \leq \tau[s,t] \quad (1.22)$$

As we wanted to see.

2. Set $-\infty<s<+\infty$ is fulfilled; τ[s, t] is not decreasing with s, in the sense that $[s`,t`] \subseteq [\tau[s,t],t]$

That *implies* $\tau[s`,t`] \geq \tau$

This is because of

$\tau[s,t] \leq s` \leq t` \leq t$

Then

$\max\{m \leq s`: U_k < a_{k-m}, \forall k \in [m,t`]\} \geq \max\{m \leq s: U_k, \forall k \in [m,t]\}$

$\max\{m \leq s`: U_k < a_{k-m}, \forall k \in [m,t`]\} \geq \max\{m \leq s: U_k < a_{k-m}, \forall k \in [m,t]\}$

then also holds that

$\max\{m \leq s: U_k < a_{k-m}, \forall k \in [m,t]\} \leq s`$

$U_k < a_{k-m}, \forall k \in [m,t]\} \leq s`$

for all

$U_k < a_{k-m}$

$k \in [m,t`]$

because

$t` \leq t$

3. Note also what do

$\{\tau|s,t| \leq j\}$

is a stop time left over U [s; t] in that *F*[S; t] · j only depends on the values

$\{U_i: i \in |j,t|\}$ (1.24)

for j≤s

Lemma 1.3.3. If $\sum_{m\geq 0} \beta_m = 0$ then for each −∞<s≤t<+∞ the random variable

$\tau[s,t]$

is an "honest" going, i.e.:

$\sum_{i\in Z} P(\tau|s,t| = i) = 1$ (1.25)

then for each 1 <s the random variable $\tau[s,+\infty]$ is honest, and

$\sum_{i\in Z} P(\tau[s,+\infty] = i) = 1$ (1.26)

$\prod_{k=0}^{\infty} a_k > 0 \quad -\infty < s$

$\leftrightarrow \forall m \in [j,s], \exists n \in [s,t]: W_n^{m-1} = 0$

$\max\{m < s: \forall n \in [s,t], W_n^m > 0\} < j-1$ "House of cards" transition matrix

$\leftrightarrow \exists n \in [s,t]: W_n^{j-1} = 0$

Wn0 where m is the nth state of the process Q: = (Q (x, y)) x; Y2E such that 8<ax

in another case

Q(X, y) =: 10 ax

beginning at time m in the state 0. It is then true that

$$\tau[s,t] = 1 + \max[m \leq s >: W\frac{o,m}{n} > 0, \forall n \in [s,t]$$

So if m · s satisfies:

$$\{\tau[s,t] < m\} \subset \cup_{i\in[s,t]}\{W_i^{o,m-1} = 0\} \tag{1.28}$$

Invariance under translations now we have to use the hypothesis and slogan A.0.3 (a) know that the $W_i^{0,0}$ chain; $m \to \infty$

$$P(\bigcup_{i\in[s,t]} W_{i-m+1}^{0,m-1} = 0\}) = P(\bigcup_{i\in[s,t]} W_{i-m+1}^{0,0} = 0\}) \leq \sum_{i=s}^{t} P(W_{i-m+1}^{0,0} = 0)$$

$$= \sum_{i=1}^{t-s} P(W_{s-m+1}^{0,0} = 0) = 0$$

It is not recurrent positive, then each of the terms

$$P(W_{s-m+1}^{0,0}) = 0$$

tends to 0 when

$$m \to \infty$$

$$\sum_{i\in Z} P(\tau[s,t] = i) = 1$$

$$P(\tau\{s,\infty] < m) \leq P(\cup_{i=s-m+1}^{\infty}\{W_i^{0,0} = 0\}) \leq \Sigma_{i=1}^{\infty} P(W_i^{0,0} = 0) \tag{1.29}$$

and therefore it holds that now for the second statement of the theme. Note that:

$$\sum_{i\in Z} P(\tau[s,t] = i) = 1$$

But as the chain is important hypothetically then each term of the right tends to zero as 1.29 m! 1. In this way:

$\sum_{i \in Z} P(\tau[s,\infty] = i) = 1$

We know from above that results:

$[0, a_k) \subset \cup_{l=0}^{k} B_t(\underline{w}) = [0, \sum_{y \in E} a_k(y; \underline{w}))$ $\quad \forall \underline{w} \in E^{-N^*}$ (1.30)

As a result we have:

$[0, a_k) \cap B_t(\underline{w}) = \emptyset \qquad \forall l > k$ (1.31)

$B_{l;k}(w_{-1}, \ldots, w_{-k}) := [0, a_k) \cap B_t(\underline{w})$ (1.32)

$B_{l,k}(y; w_{-1}, \ldots, w_{-k}) := [0, a_k) \cap B_t(\underline{w})$ (1.33)

that makes sense for all 1.31 l; k ¸ 0. Similarly to 1.21 for U 2 [0; 1] we define

Z (1.34)

Proposition 1.3.4. Be y_0 one element E. arbitrarily define the function $[0,1]^Z$ such that $\emptyset(\underline{u}) = \underline{x}$ recursively for $j = \tau[n](u)$ until $j = n$ as follows: noticed

$j = \tau[n](u)$

and define:

$$x_r := \sum_{y \in E} y \; 1_{\{u_r \in B_0(y)\}}$$

$$x_{r+1} := \sum_{y \in E} y \; 1_{\{u_{r+1} \in B_0(y) \cup B_{1;1}(y; x_r)\}}$$

$$x_n := \sum_{y \in E} y \; 1_{\{u_n \in S_{j=0}^{n-r} B_{j;n-r}(y; x_{n-1}, \ldots, x_r)\}} \quad (1.35)$$

what if $\tau[n](u)$ is finite, and $x_n = x_0$ what if $\tau[n](u) = -\infty$ (note that each sum in the above definition is reduced to a single term). So

1. Φ is well defined and measurable.

2. For each $n \in Z$ Nth component $\emptyset(\underline{u})$ we note that $\emptyset(\underline{u})_n$ depends only on the values u_i in the interval $[n](u)$what if $\tau[n](u) > -\infty$ and went: i$\{u_i : i < n\}$ otherwise.

3. $A_n := \{\underline{u} \in [0,1]^Z : \tau[n](\underline{u} \geq -\infty)\}$

then the law Φ (U) is compliant with the P.

Proof. Let $A_n := \{\underline{u} \in [0,1]^Z : \tau[n](\underline{u} \geq -\infty)\}$

1. An assembly in the consistency of the definition 1.35 is due to the following facts:

by definition,

(i) $\tau[n] \leq \tau[j]$ for all $j \in [\tau[n], n]$

(ii) And $U_n < a_k \quad x_{n-1}, \ldots, x_{n-k}$

as worth 1.30 then this shows that the event

$F(U_i : i \leq n)$

is fulfilled, then just look at the values $x_{n-1}, \ldots, x_{n-k}$ to obtain the value of x_n. These two facts imply that for all k_2 values

$\emptyset(\underline{u}) = (x_n : n \in Z)$

(iii) value x_k calculated using 1.35 with k instead of n is the same value as that obtained as part of the recursive computation for x 1.35_n. The F(u_i I · n) x – it is directly due to the definition 1.35. As ensembles, they are F (U_i I · n) -measurable maps x_n continue to be measurable after "them together." We conclude that the map Φ (u) = (x_n : N 2 Z) is well defined and measurable.

2. It follows directly from the definition of the function.

3. When An is fulfilled, 1.31 means that 1.35 is

$x_n = \sum_{y \in E} y \, 1_{\{u_n \in S_{k \geq 0} B_k(y; x_{n-1}, \ldots, x_{n-k})\}}$

And this implies that

$$P(\emptyset(\underline{U})_n = y \mid U_i : i < n; A_n) = P(U_n \in$$

$$\cup_{k \geq 0} B_k(y; (\emptyset(\underline{U})_{n-1}, \ldots, (\emptyset(\underline{U})_{n-k}) \mid U_i : i < n; A_n \quad (1.37)$$

As Φ (Uj) is F (Ui: i · j) -measurable and each A is independent of F (Ui: i · n), if A_n has total measurement time is equal to 1.37

In other words:

$$\bigcup_{k \geq 0} B_k(y; (\emptyset(\underline{U})_{n-1}, \ldots, (\emptyset(\underline{U})_{n-k}| = \sum_{k \geq 0} |B_k(y; \emptyset(\underline{U})_{n-1}, \ldots, \emptyset(\underline{U})_{n-k})|$$

$$= P(y; (\emptyset(\underline{U})_{n-1}, \ldots, (\emptyset(\underline{U})_{n-k})$$

$$P\left(\emptyset(\underline{U})_n = y \mid U_i : i < n\right) = P(y; \emptyset(\underline{U})_i : i < n) \quad (1.38)$$

As the right side of 1.38 only depends on

$\emptyset(\underline{U})_i \quad i < n$

then the left side also depends only on those values, therefore the law Φ (U) n given

$\{\emptyset(\underline{U})_i : \quad i < n$

remains given by P, then the distribution of $\emptyset(\underline{U})_i$ is compatible with P.

CHAPTER 2

POISSON PROCESSES

CONTENTS

In this chapter, we begin with a constructive definition of a Poisson process. In the fourth section we will see some properties of the process that will help us to test in Section 2.6 that the constructed process verifies the classical Poisson process definitions. In Sections 2.7 and 2.8 define Poisson processes two-dimensional and see how to build a Poisson process one-dimensional from one two-dimensional. In the last two sections we introduce the notion of superposition of Poisson processes and inhomogeneous processes.

2.1. PUNCTUAL PROCESSES

Let E be a subset of a Euclidean space (for us will be sufficient to assume that E is contained in $R^d, d \geq 1$. We randomize points in the set E and to have a convenient notation for the function which counts the number of random points that fall into a subset Dimensioned μ E.

Set

$\{X_n\}_{n \in N}$

to random elements in the array E that represent points state space E

$X_n: \Omega \to E, \forall n \in N)$

We define

$e_{X_n}: \Omega \times P(E) \to \{0,1\}$ as follows:

$$e_{X_n}(w, A) := e_{X_n(w)}(A) = \begin{cases} 1 & if \quad X_n(w) \in A \\ 0 & if \quad X_n(w) \notin A \end{cases}$$

where $P(E) = \{A \subset E\}$, e.g., is the set of subsets of $E = [0, \infty) \times R^2$, we define the counting measure $N: \Omega \times P(E) \to N$.

Thus, $N: \Omega \times P(E) \to N$ is a random variable that counts the number of points falling in A. We call $N: \Omega \times P(E) \to N$ point process and $\{X_n\}_{n \in N}$ the set of adequate points.

Observation 2.1.1. To facilitate the notation will remove the dependence of the process it has to.

Example 2.1.2. Suppose we want to model the time and location where the this case presents a possible choice space states could be $E = [0, \infty) \times R^2$ and the point process could be represented by:

$$N(\cdot) = \sum_{n \in N} e_{\left(T_n,\left(L_{n_1},L_{n_2}\right)\right)}(\cdot)$$

where

T_n *represents the time the Nth earthquake and* (L_{n_1}, L_{n_2}) represent the latitude and longitude where the nth earthquake occurs.

For t> 0 and $B \subset R^2, N([0,t] \times B)$could be the number of earthquakes up to time t in the region B. If we wanted to be more precise in identifying earthquakes, we might add, for example, the intensity with which this occurs. In this case the state space could be

$[0, \infty) \times R^2 \times [0, \infty)$

and the point process would X and

$$\tilde{N}(\cdot) = \sum_{n \in N} e_{\left(T_n,\left(L_{n_1},L_{n_2}\right),I_n\right)}(\cdot)$$

where indicates the intensity of the nth earthquake.

Note that the notation is flexible enough to make smooth changes if we want to model processes with multiple points in one place. For example, consider as before fX$_n$g$_{n2N}$ a random sequence of elements E (which might now represent "sites"), and is f³$_n$g$_{n2N}$ a sequence of non-negative integer values representing the random number of dots in random site X$_n$. Then the point process with multiple points on their sites can be represented as:

$$N(\cdot) = \sum_{n \in N} \xi_n e_{X_n}(\cdot)$$

Then for a region we have ½ E

$$\sum_{n \in N} \xi_n e_{X_n}([0,t])$$

represents the number of points in A, since we are counting the number of points on each site A.

Consider an example. Suppose that a certain place arrive bus according to a point process with fXng points, i.e., the nth bus arrives in time Xn. If ³n represents the number of passengers arriving in the Nth bus, total passengers arriving at time t is:

Definition 2.1.3. Let

$\mu(A) := P(E) \to R$

be one-time process. We define the mean or *intensity of said process as*

$\mu(A) := P(E) \to R$

such that R

$\mu(A) := E(N(A)) \forall A \subset P(E)$

In other words '(A) is the expected number of points in the region A.

2.2. POISSON PROCESSES DEFINITION

N is a point process with state space E. Let E ½ P (E) in one algebra ¾-E, i.e.,

1. $\mathcal{E} \neq \phi$
2. If then $A \in \mathcal{E}$ $A^c \in \mathcal{E}$

If $\{A_n\}_{n\in N} \subset \mathcal{E}$ then $\cup_{n\in N} A_n \in \mathcal{E}$.

Definition 2.2.1. We say that the N process is a Poisson process with mean [1] if it satisfies the following properties:

1. *For all A 2 E it is true:*

$$P(N(A) = k) = \begin{cases} \frac{\mu(A)^k}{k!} e^{-\mu(A)} & \mu(A) < \infty \\ 0 & \text{in other case} \end{cases}$$

2. *If* $A_1, .. A_k$ *are subsets of disjoint E in E, then* $N(A_1), \dots, N(A_k)$: $\Omega \to N$

are independent random variables.

In other words N is a Poisson process if the random number of points in the set A, that is,

$N(A): \Omega \to N$

has distribution of Poisson with parameter

$\mu(A)$

and the number of points in disjoint regions are going independent.

Observation 2.2.2. 1. When E = R property (2) of this definition is called independent property increases, as it implies that if

$\mu(A)$

$t_1 < t_2 < \cdots < t_k$ *with*

$t_i \in R \quad \forall i = 1, \dots, k$

and

$\{N([t_i, t_{i+1}]), i = 1, \dots k - 1\}$

are independent random variables.

When the average is a multiple of the Lebesgue measure (i.e., is the length when

$E = R^2$

, if E = the area R^2, the volume if

$E = R^3$

2. Say that the process is homogeneous. Let us now see a constructive definition of a Poisson process. Let's start considering $\{A_i\}_{i \in Z} \subset P(R)$ such that $A_i = [l_i, l_{i+1})$ with $l_i < l_{i+1}, \forall i \in Z$ a partition of R. For each $i \in Z$ call

$$|A_i| := l_{i+1} - l_i$$

the length of the interval Ai, and given

$\lambda \in R,\ \lambda > 0$ we assign each a random variable interval $Y_i: \Omega \to N$ with distribution of Poisson with parameter $\lambda |A_i|$ ie:

$$P\left(Y_i = k\right) = \frac{(\lambda \mid A_i \mid)^k}{k!} e^{-\lambda |A_i|}$$

we assume;

$\{Y_i, i \in Z\}$

is a family of independent random variables.

Second, we assign each

$i \in Z$

a succession of $U_{i,j}$ where $\{U_{i,j} \in N^*\}$ variable, j with j 2 N have uniform distribution in the interval A_i i.e:

$$P(U_{i,j} \in A \cap A_i) = \frac{|A \cap A_i|}{|A_i|}, \quad \forall A \subset R$$

We define $S: \Omega \to R$ the random set given by:

$$S := \bigcup_{i \in z} S_i$$

where

$\{\widetilde{S_1}\}_{n \in Z}$

is defined by:

$$S_i = \begin{cases} \{U_{i,j}: 1 \le j \le Y_i\} & if \quad Y_i \ge 1 \\ \phi \quad if & Y_i = 0 \end{cases}$$

Finally let Si be orderly succession of points S, where it is the first point of this S to the right of the origin and

$$\tilde{S}_n = \begin{cases} \min\{s > \tilde{S}_{n-1}: s \in S\} & if \quad n \geq 2 \\ \max\{s < \tilde{S}_{n+1}: s \in S\} & if \ n \leq 0 \end{cases}$$

$$N_S(w, A) := \sum_{n \in N} 1_{\{S_n(w) \in A\}}$$

$A \subset R$ for all R, then $S \cap A$ is the number of points in the set $N(A)$. When not confusingly write N (A) instead of N_S when we east-mos referring to the random variable

$N_S(A): \Omega \rightarrow N$

defined above process will call Poisson process.

2.3. DEFINITION FORM OF A POINT PROCESS

In this section we discuss in more detail the state space of our model and see that there are certain sets in the space that determine the distribution of a point process.

Let E be the space where the points of our model live (in our case E will be a subset of

R^d

with

$d \geq 1$

as stated in the first section of this chapter) and assume that E is a ¾-algebra in E (for example we consider ¾-algebra generated by the open sets E or equivalently, the generated the "rectangles" E). One way to model a random distribution of points in E is to fX$_n$ g$_{n2}$ N random elements E and then say that the stochastic point process is the counting function that the region

$A \in \mathcal{E}$

assigns the number of random elements fX$_n$g they are in A, as we did in the first section of this chapter. Below repeat that building a more formal way. Given $x \in E$ and 2 E defined as delta follows:

A point measure E is a measure of how

$$e_x(A) := \begin{cases} 1 & if \quad x \in A \\ 0 & if \quad x \notin A \end{cases}$$

where xi is in E and all bounded region A contains only a finite number of points xi. Then m (A) is the number of xi that are in her region A. Denote by $M_p := M_p(E)$, the set of all specific measures on E and consider subsets of the form Mp

$$m := \sum_i e_{x_i}$$

where k≥0 and R is a rectangle bounded in E. We call

$$\{m \in M_p : m(I) = k\}$$

to lower ¾-algebra containing said sets (where I can vary). We can now define a point process as a random element (M_p; M_p), That is to say, given a space probability (Ω; F; P) where F is the ¾-algebra of events, N is a timely process if a measurable map

$$N: (\Omega, F) \to (M_p, M_p)$$

Which means that for all

$$\Lambda \in M_p$$

It is true that

$$N^{-1}(\Lambda \in M_p, M_p)$$

In particular, if N is a timely process then

$$\{N(I) = k\} = N^{-1}(\{w \in M_p : m(I) = k\}) \in F$$

$$N^{-1}(\Lambda) := \{w \in \Omega : N(w) \in \Lambda\} \in F$$

as

$$\{m \in M_p : m(I) = k\} \in M_p$$

Now consider a point process:

$$N: (\Omega, F) \to (M_p, M_p)$$

and a probability measure P in (Ω, F). The distribution P N is the measure

$$P \circ N^{-1} = P(N \in (\cdot)) \qquad (M_p, M_p).$$

Then the probability of an event dependent on N can be determined if we know the distribution of N. define finite dimensional distributions of N as the collection of functions:

$$p_{I_1,\dots I_k}(n_1, \dots n_k) := P(N(I_1) = n_1, \dots, N(I_k) = n_k$$

where $I_1, \dots, I_k$ are rectangles bounded and

$$n_1, \dots, n_k$$

are non-negative integers.

Theorem 2.3.1. Finite dimensional distributions of a point process N determined in a unique way the distribution

$$P \circ N^{-1}$$

Proof. Given

k≥0

and R a rectangle bounded in E consider sets

$E_{k,I} := \{m \in M_p: m(I) = k\}$

G is the class of finite intersections of sets of the form 2.4

$G := \{\cap_{i=1}^{n} E_{k_i,I_i}: k_1, \dots, k_n \in N, I_k$ rectangle bounded $\forall i = 1, \dots n, n \in N\}$

Then as G is closed under finite intersections and generates the ¾-algebra Mp must be a measure Mp is only determined by the values taken in G (Billingsley, 1986 p. 38). In particular, $0 < p_i < 1$ is a measure defined on Mp and G values are given by finite dimensional distributions.

2.4. PROPERTIES

The purpose of this section is to prove that the construction made in Section (2.2) actually is a Poisson process, i.e., the process defined by 2.2 verifies the definition 2.2.1. We start seeing some properties of random variables with Poisson distribution that will be useful to achieve the objective.

Lemma 2.4.1. If 0 <pi <1 for all i >= 0 then:

$$\lim_{N\to\infty} \prod_{i=0}^{N} p_i = 0 \leftrightarrow \sum_{i=0}^{\infty} q_i := \sum_{i=0}^{\infty} (1 - p_i) = \infty$$

$$2. \prod_{i=0}^{\infty} p_i > 0 \leftrightarrow \sum_{i=0}^{\infty} q_i := \sum_{i=0}^{\infty} (1 - p_i) < \infty$$

Proof. We say that

$$a_n \sim b_n$$

when n

an>>bn It is true that

we can equivalently say that for every [2]

1 If

$$\lim_{n\to\infty} \frac{a_n}{b_n} = 1$$

2 . So if bn <1.

3 0 for all n has sufficiently larger than

$$(1 - \varepsilon)b_n < a_n < (1 + \varepsilon)b_n$$

Using the above and

$$-\log(1-x)\sim x$$

when x → 0then we have to

$$\prod_{i=0}^{\infty} p_i > 0 \leftrightarrow \sum_{i=0}^{\infty} -\log(1-q_i) < \infty$$

$$\leftrightarrow \sum_{i=0}^{\infty} q_i < \infty$$

Note that both

$$\sum_{i=0}^{\infty} -\log(1-q_i) < \infty$$

and ^{P1} q*i* <1 imply that

$$\sum_{i=0}^{\infty} q_i < \infty$$

Lemma 2.4.2. *1. Observe*

$$X_1, X_2, \dots, X_n$$

independent random variables with distribution R Poisson parameter;

$$\lambda_i, i = 1, \dots, n$$

respectively. It is true that *the random variable*

$$X_1 + \cdots + X_n$$

has Poisson distribution parameter

$$\lambda_1 + \cdots + \lambda_n$$

2. Consider

$$\{X_n\}_{n \in N}$$

sequence of independent random variables with distribution Poisson parameter respectively.

$$\sum_{n \in N} X_n$$

where $\sum_{n \in N} \lambda_n < \infty$ *then the variable Xn has the Poisson distribution*

*Proof.*1. We induction in the amount of n random variables. If n = 2: Sea k2

N. Using the formula of total probability we have to

$$P(X_1 + X_2 = k) = P(X_1 = k - X_2) = \sum_{k=0}^{\infty} P(X_1 = k - h, X_2 = h)$$

As X1 and X2 are independent and have distribution of Poisson, and noting that , 0 X1 have the above it is equivalent to

$$\sum_{h=0}^{k} P(X_1 = k - h)P(X_2 = h) = \sum_{h=0}^{k} \frac{\lambda_1^{k-h}}{(k-h)!} e^{-\lambda_1} \frac{\lambda_2^h}{h!} e^{-\lambda_2}$$

$$= e^{-(\lambda_1+\lambda_2)} \frac{\lambda_1^k}{k!} \sum_{h=0}^{k} C_h^k (\frac{\lambda_2}{\lambda_1})^h 1^{h-k}$$

$$= e^{-(\lambda_1+\lambda_2)} \frac{\lambda_1^k}{k!} (1 + \frac{\lambda_2}{\lambda_1})^k$$

$$= e^{-(\lambda_1+\lambda_2)} \frac{\lambda_1^k}{k!} \frac{(\lambda_1 + \lambda_2)^k}{\lambda_1^k}$$

$$= e^{-(\lambda_1+\lambda_2)} \frac{(\lambda_1 + \lambda_2)^k}{k!}$$

thus it is proved that $X_1 + X_2$ is a random variable with parameter Poisson distribution of

$$\lambda_1 + \lambda_2$$

Suppose now that the result is true for n variables. Then using the formula of total probability and independence between the variables again we have:

$$P(\sum_{i=1}^{n+1} X_i = k) = P(\sum_{i=1}^{n+1} X_i = k - X_{n+1})$$

$$\sum_{h=0}^{k} P(\sum_{i=1}^{n} X_i = k - h, X_{n+1} = h)$$

Applying the induction hypothesis and using that Xn + 1 is Poisson with parameter n + 1 we get that the above is:

$$\sum_{h=0}^{k} P(\sum_{i=1}^{n} X_i = k - h)\, P(X_{n+1} = h)$$

$$\sum_{h=0}^{k} \frac{(\sum_{i=1}^{n} \lambda_i)^{k-h}}{(k-h)!} e^{-(P_{i=1}^{n} \lambda_i)} \frac{\lambda_{n+1}^h}{h!} e^{-\lambda_{n+1}} =$$

$$e^{-(P_{i=1}^{n}\lambda_i)}e^{-\lambda_{n+1}}\sum\nolimits_{h=0}^{k}(\sum_{i=1}^{n}\lambda_i)^{k-h}\lambda_{n+1}^{h}\frac{C_h^k}{k!}=$$

$$\frac{\sum_{i=1}^{n+1}\lambda_i}{k!}e^{-(P_{i=1}^{n+1}\lambda_i)}$$

which is proved with part (1) of the lemma.

2. We define $Y_n$variables: $Y_n = X_1 + \cdots + X_n$ such that $Y_n: \Omega \to N \cup \{+\infty\}$.

Consider the set

$A_\infty = \{w \in \Omega: Y(w) = \infty\}$

and the variable

$$Y(w) = \sum_{i \in N} X_i(w)$$

where $\hat{Y}: \Omega \backslash A_\infty \to N$Note that if

$\hat{Y} = Y|_{\Omega \backslash A_\infty}$

then the variables

$Y \; and \; \hat{Y}$

Theyhavethesamedistribution.Nowconsiderforeach $k \in N$ and each $n \in N^*$ sets:

$$\Omega_{n,k} = \{w \in \Omega: \sum_{i=1}^{n} X_i(w) = kyX_m(w) = 0 \quad if \quad m > n\} \tag{2.1}$$

$$\Omega_{\infty,k} = \{w \in \Omega: \hat{Y}(w) = k\} \tag{2.2}$$

Note that

$\Omega_{n,k} \subseteq \Omega_{n+1,k}$

$\cup_{n \in N} \Omega_{n,k} = \Omega_{\infty,k}$

using part (1) the motto we have:

$$P(\Omega_{\infty,k}) = \lim_{n\to\infty} P(\Omega_{n,k})$$

$$= \lim_{n\to\infty} [\frac{e^{-P_{i\in N}\lambda_i}}{k!}(\lambda_1 + \cdots \lambda_n)^k]$$

$$= \frac{e^{-P_{i\in N}\lambda_i}}{k!} \lim_{n\to\infty} (\lambda_1 + \cdots \lambda_n)^k$$

$$= \frac{e^{-P_{i \in N} \lambda_i}}{k!} \left(\sum_{i \in N} \lambda_i\right)^k$$

where the last equality is valid for the continuity of the function

$$f(x) = x^k$$

This implies that the variable Y has Poisson distribution of parameter

$$\sum_{i \in N} \lambda_i$$

Then to prove the motto is just to prove that

$$P(A_\infty) = 0$$

$$P(A_\infty) = P(\{w \in \Omega : Y(w) = \infty\})$$

$$= P(\{w \in \Omega : X_i(w) \neq 0 \quad i \in N\})$$

for infinite sets. Consider

$$A_n := \{w \in \Omega : X_n(w) \neq 0\}$$

So:

$$\Sigma_{n \in N} P(A_n) = \Sigma_{n \in N} P(X_n \neq 0) = \Sigma_{n \in N} 1 - e^{-\lambda_n}$$

On the other hand we know that

$$\sum_{n \in N} \lambda_n < \infty$$

So

$$\prod_{n \in N} e^{-\lambda_n} = e^{-P_{i \in N} \lambda_i} > 0$$

$$\sum_{n \in N} P(A_n) < \infty$$

we conclude using the Borel-Cantelli lemma that

(A_n)=∞

thus ending the Proof of Lemma.

Proposition 2.4.3. For each interval bounded $A \subset R$ the random variable $(A){:}\Omega \to N$ has Poisson distribution with mean $\lambda|A|$..

Proof. Observe first that as the intervals Ai with $i \in Z$ are disjoint, then the intervals

$A \cap A_i$

are also. This implies that the random variables

$N(A \cap A_i)$

are independent since, given

$i \neq j$

„$k \in N$ we have:

$$P(N(A \cap A_i) = k, N(A \cap A_j) = h) = P(\sum_{n \geq 0} 1_{\{S_n \in A \cap A_i\}} = k, \quad \sum_{n \geq 0} 1_{\{S_n \in A \cap A_j\}} = h) =$$

$$P(\exists n_1, \ldots, n_k \in Z: \tilde{S}_{ni} \in A \cap A_i, \forall l = 1, \ldots k; (\exists m_1, \ldots, m_h \in Z: \tilde{S}_{mr} \in A \cap A_j, \forall r = 1, \ldots h) =$$

$$(\exists r_1, \ldots, r_k \in N: U_{i,rl} \in A \cap A_i, \forall l = 1, \ldots k; \exists s_1, \ldots, s_h \in N: U_{j,st} \in A \cap A_j, \forall t = 1, \ldots h;$$

Now considering the variables Ui; j are independent we obtain that the above is:

$$P(\sum_{r=1}^{+\infty} 1_{\{U_{i,r} \in A \cap A_i\}} = k) \quad P(\sum_{r=1}^{+\infty} 1_{\{U_{j,r} \in A \cap A_j\}} = h) =$$

$$P(N(A \cap A_i) = k) \quad P(N(A \cap A_j) = h)$$

Then these are independent variables.

Let k in N partitions the space according to the possible values of the variable-Yi we have: $(N(A \cap A_i)=k)$ $P(N(A \cap A_j)=h)$

Now notice that if $h<k$ $Y_i=h$ then in

$S \cap (A \cap A_i)$

will be at most Yi = h points (for the definition of Si) and therefore

$N(A \cap A_i)$

can not take the k value. 2.2 then becomes:

$$\sum_{h=k}^{+\infty} P\left(Y_i = h, \quad N(A \cap A_i) = k = \right.$$

$$\left. \sum_{h=k}^{+\infty} P(Y_i = h, \quad \sum_{j=1}^{+\infty} 1_{\{U_{j,r} \in A \cap A_j\}} = k\right) \tag{2.3}$$

Given that only fall into the set S $U_{i,}$ and that the variables $U_{i,j}$ They are independent of Y_i 2.9 is:

$1 \leq j \leq Y_i$

Assertion (1): is a variable with Bernoulli distribution with parameter

Proof of the claim (1):

Since the variables $U_{i,j}$ are uniform in the range A_i. the variables are independent since the $U_{i,j}$ are equal to:

$$\sum_{h=k}^{+\infty} P\left(Y_i = h, \sum_{j=1}^{+\infty} 1_{\{U_{i,j} \in A \cap A_i\}} = k\right) =$$

$$\sum_{h=k}^{+\infty} P\left(Y_i = h\right) \sum_{j=1}^{+\infty} 1_{\{U_{i,j} \in A \cap A_i\}} = k \Big) \qquad (2.4)$$

Affirmation :

$$\sum_{j=1}^{h} 1_{\{U_{i,j} \in A \cap A_i\}})$$

with parameters $\frac{|A \cap A_i|}{|A_i|} 1_{\{U_{i,j} \in A \cap A_i\}}$ is a random variable with binomial distribution

$$\{1_{\{U_{i,j} \in A \cap A_i\}}\}_{j \in N^*}$$

Proof of the claim (2):

$$P(\sum_{j=1}^{h} 1_{\{U_{i,j} \in A \cap A_i\}} = k) =$$

$$P(\exists j_1, \dots j_k) \in \{1, \dots, h\}: U_{i,ji} \in A \cap A_i, \quad \forall l = 1, \dots, k, \; U_{i,j} \notin A \cap A_i \; if \; j \in N^* \setminus \{j_1, \dots, j_k\})$$

Then, once elected

$\{j_1, \dots, j_k\} \in N^*$

we have that the following is valid:

$$P(U_{i,ji} \in A \cap A_i, \quad \forall l = 1, \dots, k, \; U_{i,j} \notin A \cap A_i \; if \; j \in N^* \setminus \{j_1, \dots, j_k\}) =$$

$$\prod_{l=1}^{k} P(U_{i,ji} \in A \cap A_i)$$

$$\prod_{j \in \{1,\dots,h\} \setminus \{j_1,\dots,j_k\}} P(U_{i,j} \notin A \cap A_i) =$$

$$(\prod_{l=1}^{k} \frac{|A \cap A_i|}{|A_i|})(\prod_{j \in \{1,\dots,h\} \setminus \{j_1,\dots,j_k\}} (1 - \frac{|A \cap A_i|}{|A_i|})) =$$

$$(\prod_{l=1}^{k} \frac{|A \cap A_i|}{|A_i|})^k (1 - \frac{|A \cap A_i|}{|A_i|})^{h-k}$$

To conclude with the statement that we observe

$$C_k^h$$

different ways to choose $J_1,\ldots,J_k$ therefore:

$$P(\sum_{j=1}^{h} 1_{\{U_{i,j}\in A\cap A_i\}} = k) =$$

$$P(\exists j_1, \ldots j_k) \in \{1,\ldots,h\}: U_{i,ji} \in A\cap A_i, \quad \forall l = 1,\ldots,k, \ U_{i,j} \notin A\cap A_i \ if\ j\in N^* \setminus \{j_1,\ldots,j_k\})$$

$$P(U_{i,ji} \in A\cap A_i, \quad \forall l = 1,\ldots,k, \ U_{i,j} \notin A\cap A_i \ if\ j\in N^* \setminus \{j_1,\ldots,j_k\}) =$$

$$\prod_{l=1}^{k} P(U_{i,ji} \in A\cap A_i) \quad \prod_{j\in\{1,\ldots,h\}\setminus\{j_1,\ldots,j_k\}} P(U_{i,j} \notin A\cap A_i) =$$

whereby the assertion is proved (2).
Applying now the claim (2) 2.10 becomes:

$$(\prod_{l=1}^{k} \frac{|A\cap A_i|}{|A_i|})(\prod_{j\in\{1,\ldots,h\}\setminus\{j_1,\ldots,j_k\}} (1-\frac{|A\cap A_i|}{|A_i|})) =$$

$$(\prod_{l=1}^{k} \frac{|A\cap A_i|}{|A_i|})^k (1-\frac{|A\cap A_i|}{|A_i|})^{h-k}$$

Then

$$P(\sum_{j=1}^{h} 1_{\{U_{i,j}\in A\cap A_i\}} = k) = C_k^h (\prod_{l=1}^{k} \frac{|A\cap A_i|}{|A_i|})^k (1-\frac{|A\cap A_i|}{|A_i|})^{h-k}$$

$$\sum_{h=k}^{+\infty} P(Y_i = h)\, C_k^h (\frac{|A\cap A_i|}{|A_i|})^k (1-\frac{|A\cap A_i|}{|A_i|})^{h-k}$$

$$\sum_{h=k}^{+\infty} \frac{(\lambda|A_i|)^h}{h!} e^{-\lambda|A_i|} \frac{h!}{(h-k)!\,k!} (\frac{|A\cap A_i|}{|A_i|})^k (1-\frac{|A\cap A_i|}{|A_i|})^{h-k}$$

$$= \frac{e^{-\lambda|A_i|}}{k!} (\frac{|A\cap A_i|}{|A_i|})^k \lambda^k |A_i|^k \sum_{u=0}^{+\infty} \frac{\lambda^u |A_i|^u}{u!} (1-\frac{|A\cap A_i|}{|A_i|})^u$$

$$= \frac{e^{-\lambda|A_i|}}{k!} (\frac{|A\cap A_i|}{|A_i|})^k \lambda^k |A_i|^k \sum_{u=0}^{+\infty} \frac{\lambda^u |A_i|^u}{u!} (1-\frac{|A\cap A_i|}{|A_i|})^u$$

$$= e^{-\lambda|A_i|} \frac{|A\cap A_i|^k}{k!} \lambda^k e^{\lambda(|A_i| - |A\cap A_i|)}$$

$$= \frac{(\lambda|A\cap A_i|)^k}{k!} e^{-\lambda|A\cap A_i|}$$

Finally note that $N(A\cap A_i) \sim Poisson(\lambda|A\cap A_i|) \forall i \in Z$ then the sets A_i are disjoint, therefore

$N(A) = \sum_{i \in Z} N(A \cap A_i)$ is the sum of

$\bigcup_{i \in Z} A_i = R$ and therefore

$\bigcup_{i \in Z}(A \cap A_i) = A$ applying $\sum_{i \in Z} |A \cap A_i| = |A|$

$N(A) \sim Poisson(\lambda |A|)$

Proposition 2.4.4. For each finite family of disjoint interval$B_{1},\ldots,_{L}$ random variables

$\{N(B_i)\}_{i=1}^{L}$

are independent and have Poisson distribution with media

$\lambda |B_l|, l = 1, \ldots L$

respectively.

Proof. Let $i \in Z$ and see that given

$$N(A_i) = h_i$$

$$\{N(B_l \cap A_i): l = 1, \ldots, L\}$$

have multinomial distribution, i.e., for integer values

$$k_{l,i} \geq 0$$

$$\sum_{l=1}^{L} k_{l,i} = h_i$$

$$b_{l,i} \geq 0$$

$$\sum_{l=1}^{L} b_{l,i} = 1$$

such that

$$P\big(N(B_t \cap A_i) = k_{l,i}, l = 1, \ldots, L | N(A_i) = h_i\big) = h_i! \frac{\prod_{l=1}^{L} b_{l,i}^{k_{l,i}}}{\prod_{l=1}^{L} k_{l,i}!}$$

and consider

$B_{L+1}: (\cup_{l=1}^{L} B_l)^c$ (2.13)

$b_{l,i} := \frac{|B_l \cap A_i|}{|A_i|}, \forall 1 \leq l \leq L$

$$b_{L+1,i} := 1 - \sum_{l=1}^{L} b_{l,i}$$

$k_{l,i} \geq 0, l = 1, \ldots, L+1$

$$\sum_{l=1}^{L+1} b_{l,i}$$

$$P\big(N(B_i \cap A_i) = k_{l,i}, l = 1, \ldots, L+1 | N(A_i) = h_i\big) =$$

$$\frac{P\big(N(B_i \cap A_i) = k_{l,i}, l = 1, \ldots, L+1 | N(A_i) = h_i\big)}{P(N(A_i) = h_i)}$$

Note that $\cup_{l=1}^{L+1} B_l \cap A_i = (\cup_{l=1}^{L} B_l \cap A_i) \cup (B_{L+1} \cap A_i) = A_i$

Because kl, 0; l = 1,…n arbitrary integers such that B_{L+1}

so where the latter equality is the definition of

$N(B_l \cap A_i) = k_{l,i},\ \ l = 1, \ldots, L+1$

$$N(A_i) = \sum_{l=1}^{L+1} N(B_l \cap A_i) = \sum_{l=1}^{L+1} k_{l,i} = h_i$$

Because we have

$$\frac{P(N(B_l \cap A_i) = k_{l,i},\ \ l = 1, \ldots, L+1)}{P(N(A_i) = h_i)}$$

$$\{B_l \cap A_i\}_{l=1,\ldots,L+1}$$

$$\{N(B_l \cap A_i)\}_{l=1,\ldots,L+1}$$

$$\prod_{l=1}^{L+1} \frac{P(N(B_l \cap A_i) = k_{l,i}}{P(N(A_i) = h_i)} = \frac{h_i!}{e^{-\lambda|A_i|}(\lambda|A_i|)^{h_i}} \prod_{l=1}^{L+1} e^{-\lambda||B_l \cap A_i|} \frac{(\lambda|B_l \cap A_i|)^{k_{l,i}}}{k_{l,i}!}$$

$$= \frac{h_i!}{\lambda^{h_i} |A_i|^{h_i}} e^{\lambda |A_i|} e^{-\lambda P_i^{L+1} |B_l \cap A_i|} \lambda^{P_{i=1}^{L+1} k_{l,i}} \prod_{l=1}^{L+1} \frac{|B_l \cap A_i|^{k_{l,i}}}{k_{l,i!}}$$

$$= \frac{h_i!}{\prod_{l=1}^{L+1} k_{l,i}}! \prod_{l=1}^{L+1} b_{l,i}^{kl,i}$$

As the intervals

$\{N(B_l \cap A_i)\ l = 1, \dots, L, \quad i \in Z\}$

are disjoint is that the random variables

$N(B_l \cap A_i) \sim Poisson(\lambda |B_l \cap A_i|)$

are independent, then considering this proposal and 2.4.3 turns out that 2.18 is:

$$P\big(N(B_l) = n_l, \qquad l = 1, \dots, L \mid N(A) = n\big) = \frac{n!}{n_1! \dots n_L!} \prod_{l=1}^{L} (\frac{|B_l|}{|A|})^{n_l}$$

where

$N(A) = n$

$B_1, \dots, B_l$

$n_1, \dots, n_L$

$n_1 + n_2 + \dots + n_L = n$

as we wanted to prove. Then, as the new variables (A_i) g_{iinZ} are independent to

$\{N(B_l \cap A_i)\ l = 1, \dots, L, \quad i \in Z\}$

are independent variables and

$N(B_l \cap A_i) \sim Poisson(\lambda |B_l \cap A_i|)$

Finally is noted that since

$l = 1, \dots, L$

verified that

$N(B_t) \sim Poisson\big(\lambda |B_l|\big) \forall l = 1, \dots, L$

Proposition 2.4.5. A range for all the conditional distribution of the points in $S \cap A$,$(A)=n$ is the same as that of n vai uniformly distributed in the interval A.

Proof. We will use Theorem 2.3.1. Observe

$B_1, \ldots, B_l$

a partition of A and are $n_1, \ldots, n_L$ nonnegative integers such that

$n_1 + n_2 + \cdots + n_L = n$

By 2.12 we know that:

$$P(N(B_l) = n_l, \quad l = 1, \ldots, L \mid N(A) = n) = \frac{n!}{n_1! \ldots n_L!} \prod_{l=1}^{L} (\frac{|B_l|}{|A|})^{n_l}$$

Then, by Theorem 2.3.1 it suffices to prove that if

$U_1, \ldots, U_n$ are uniformly

$$P(M(B_l) = n_l, l = 1, \ldots, L) = \frac{n!}{n_1! \ldots n_L!} \prod_{l=1}^{L} (\frac{|B_l|}{|A|})^{n_l}$$

But considering that the random variables M(Bl) are going binomial with parameters n

$\frac{|B_l|}{|A|}$

(this statement (2) made in Proposition 2.4.3 should) we have:

$$P(M(B_l) = n_l : l = 1, \ldots L) =$$

$$C_{n_1}^{n} P(U_{n_i} \in B_1 \ \forall i = 1, \ldots, n_1) \cdot C_{n_2}^{n-n_1} P(U_{n_i} \in B_2 \ \forall i = n_1 + 1, \ldots, n_1 + n_2) \cdots$$

$$\cdots C_{n_L}^{n-(n_1+n_2 \ldots + n_{L-1})} P(U_{n_i} \in B_L \ \forall i = n_1 + \cdots + n_{L-1}, \ldots, n) =$$

$$\frac{n!}{(n-n_1)!\, n_1!} (\frac{|B_1|}{|A|})^{n_1} \cdot \frac{(n-n_1)!}{(n-n_1-n_2)!\, n_2!} (\frac{|B_2|}{|A|})^{n_2} \cdots$$

$$\cdots \frac{(n-n_1-\cdots-n_{L-1})!}{n_L!} (\frac{|B_1|}{|A|})^{n_1} \cdot \frac{(n-n_1)!}{(n-n_1-n_2) n_2!} (\frac{|B_2|}{|A|})^{n_2} \cdots$$

$$\cdots \frac{(n-n_1-\cdots-n_{L-1})!}{n_L!} \left(\frac{|B_L|}{|A|}\right)^{n_L} =$$

$$\frac{n!}{n_1!\, n_2! \ldots n_L!} \prod_{l=1}^{L} (\frac{|B_l|}{|A|})^{n_l}$$

Observation 2.4.6. Propositions 2.4.4 and 2.4.5 follows that the construction of the process 2.2 not depend on the choice of Ai intervals forming the partition R.

2.5. MAIN RESULTS

This section presents an alternative construction of a Poisson process one-dimensional then we use to test certain results about the distribution of time between successive events of a Poisson process.

T1 be a random variable with exponential distribution of parameter , and N1 (¢) a Poisson process with parameter , T1 independent variable. We define N (¢) process:

$$N(A) := 1_{\{T_i \in A\}} + N_1(A - T_1)$$

for

$$A \subset R$$

$$A - t = \{x \in R : x + t \in A\}$$

In other words, the N process (¢) is obtained by setting the first event according to going T1 and adding after that instant a Poisson process Independent.

The following result ensures that the defined process before checks the properties of the Poisson process.

Theorem 2.5.1. The process 2.19 is defined by a Poisson process.

Proof. By Theorem 2.3.1 is sufficient to see that the process verifies the proposition 2.19 2.4.4. Therefore we need to calculate:

$$P(N(B_l) = k_l : l = 1, \dots L)$$

where $B_1, \dots, B_L$ are arbitrary intervals in R,

$$k_1, \dots k_L$$

nonzero integer values, $L \geq 1$. To facilitate the notation, we will test if L = 1 and B1 = [a; c] and left as a generalization to the case L , 1.

Conditioning the process value to the values of the variable T_1 we have to:

$$P(N([a,c]) = 0, T_1 < a) = \int_0^a \lambda e^{-\lambda t} P(N_1([a-t, c-t]) = 0) dt$$

$$= \int_0^a \lambda e^{-\lambda(c-a+t)} dt$$

$$= e^{-\lambda(c-a)}(1 - e^{-\lambda a}) \qquad (2.6)$$

Consider first the case k = 0. In this case the terms are to be 2.6:

$$P(N([a,c]) = 0, T_1 = [a,c]) = 0$$

as for the definition of N (¢) process that

$T_1 = [a,c]$

$N([a,c]) \geq 1$

Finally we have

$$P(N([a,c]) = 0, T_1 > 0, T_1 > c) = \int_c^{\infty} \lambda e^{-\lambda t} dt$$

$$= e^{-\lambda t}$$

because of

$T_1 > c$

$N_1([a-t, c-t]) = 0$

Then if k = 0 we get by adding 2.22, 2.23 and 2.24 that:

$$P(N([a,c]) = k) = T_1 < a) = \int_0^a \lambda e^{-\lambda t} P(N_1([a-t, c-t]) = k) dt$$

as we wanted to see.

Consider now the case k> 0. By studying each case, similar to what was done for the case k = 0 we have:

$$P(N([a,c]) = k) = T_1 < a) = \int_0^a \lambda e^{-\lambda t} P(N_1([a-t, c-t]) = k) dt$$

$$= \int_0^a \lambda e^{-\lambda t} e^{-\lambda(c-a)} \frac{\lambda^k (c-a)^k}{k!} dt$$

$$\frac{\lambda^k (c-a)^k}{k!} e^{-\lambda(c-a)} (1 - e^{-\lambda a})$$

To calculate the value of the second term of the right of 2.21, note that as

$T_1 \in [a,c]$

$a - t \leq 0$

$t \in [a,c]$

and then:

$$P(N([a,c]) = k, T_1 \in [a,c]) = \int_c^a \lambda e^{-\lambda t} P(N_1([0, c-t]) = k-1) dt$$

$$= \int_a^c \lambda e^{-\lambda t} e^{-\lambda(c-t)} \frac{(\lambda(c-t))^{k-1}}{(k-1)!} dt$$

$$= \frac{\lambda^k}{k!} e^{-\lambda c} (c-a)$$

Finally, the last term vanishes because as T1> c then

$[a-t, c-t] \subset (-\infty, 0)$

$t \geq c$

In this way we conclude

$$P(N([a,c]) = k) = \frac{\lambda^k (c-a)^k}{k!} e^{-\lambda(c-a)}$$

for all $k \in N$, which proved theorem.

Corollary 2.5.2. Sea N (¢) a Poisson process one-dimensional parameter. Consider

$S_1, S_2, \ldots$

as ordered time of occurrence of process events. Then the random variables

$T_n := S_n - S_{n-1}, n \geq 2$

$N_1(\cdot)$

are independent and identically distributed exponential-mind parameter distribution ,.

Proof. Theorem 2.5.1 tells us that a Poisson process N_0(¢) can be constructed by placing the time of the first event according to an exponential going T_1 and then a Poisson process N_1(¢) independent of T_1. Similarly, the process N_1(¢) can be built with an exponential value T and a Poisson process N_2(¢) independent. By iterating this construction, we have moments between successive events given as a series

$T_1, T_2, \ldots$

which are exponentially distributed.

2.6. DETERMINING THE ORDERING

As we saw, a way to build a Poisson process in a region bounded A is "splash" a number of Poisson uniform on A. Then the condition that if the region have n points distribution, these points are uniformly distributed in A. in this section we shall assume that the state space E is [0; +1) and we see some results valid in this case.

Definition 2.6.1. Let

$X_1, \ldots, X_n$

be a space Ω. We define new vari*ables random* $X_{(1)}$; …, $X_{(N)}$ *proficient* Ω we will call order statistics *the variables* $X_1; \ldots, X_n$ *as follows:*

$$X_{(1)}(w) := \min\{X_1(w), \ldots, X_n(w)\}$$

$$X_{(i)}(w) := \min(\{X_1(w), \ldots, X_n(w)\} \setminus \{X_1(w), \ldots, X_{i-1}(w)\}) \;\; if\ i = 2, \ldots n$$

We will now see that the joint density of the order is uniform in the range $(0, t) \subset R$

Lemma 2.6.2. Observe

$U_1, U_2, \ldots, U_n$

with uniform distribution in the interval (0, t), and their

$U_{(1)}, U_{(2)}, \ldots, U_{(n)}$

associated order. Then the joint density of the order statistics is:

$$f_{U_{(1)},U_{(2)},\ldots U_{(n)}}(u_1, \ldots, U_n) = \begin{cases} \frac{n!}{t^n} & 0 < u_1 < \cdots < u_n < t \\ & 0 \end{cases}$$

Proof. Let a function of a family

$\prod := \{f: 1, \ldots, n\} \to \{1, \ldots, n\}$

denote permutations of n elements. If

$$\pi \in \boldsymbol{\Pi}$$

$$(U_{(1)}, U_{(2)}, \ldots U_{(n)}) = (U_{\pi(1)}, U_{\pi(2)}, \ldots U_{\pi(n)})$$

$$\{U_{\pi(1)}, U_{\pi(2)}, \ldots U_{\pi(n)}\}$$

$$\{U_{\pi(1)} < U_{\pi(2)}, \ldots < U_{\pi(n)}\}$$

then for each bounded function $g(u_1, \ldots, u_n)$ we have to

$$E\left(g(U_{(1)}, U_{(2)}, \ldots U_{(n)})\right) = \sum\nolimits_{\pi \in \Pi} E(g(U_{\pi(1)} < U_{\pi(2)}, \ldots < U_{\pi(n)}) 1_{\{U_{\pi(1)} < U_{\pi(2)}, \ldots < U_{\pi(n)}\}}$$

As the joint density of

$U_{\pi(1)}, U_{\pi(2)}, \ldots U_{\pi(n)}$

is equal to

$$f_{U_{\pi(1)},U_{\pi(2)},\ldots U_{\pi(n)}}(u_1, \ldots, u_n) = f_{U_1,\ldots U_n}(u_1, \ldots, u_n)$$

and have a uniform distribution

$$f_{U_1,\dots U_n}(u_1,\dots,u_n)=\begin{cases}\frac{1}{t^n} & (u_1,\dots,u_n)\in[0,1]^n\\ & 0\end{cases}$$

we have

$$E\left(g\left(U_{(1)},U_{(2)},\dots U_{(n)}\right)\right)=\sum_{\pi\in\Pi}\int_{\{0<u_1<\dots<u_n<t\}}^{\infty} g(u_1,\dots,u_n)\frac{1}{t^n}du_1\dots du_n$$

$$=\int_{[0,1]^n}^{\infty} g(u_1,\dots,u_n)(\frac{n!}{t^n}1_{\{n_1<\dots<u_n\}}(u_1,\dots,u_n))du_1\dots du_n$$

which shows that the density of

$$\left(U_{(1)},U_{(2)},\dots U_{(n)}\right)$$

is as wanted to try

$$\frac{n!}{t^n}1_{\{n_1<\dots<u_n\}}(u_1,\dots,u_n)$$

We prove now that a homogeneous Poisson process on [0; +∞)) verifies the property of the order statistics, which means, conditioned to have n points (0, t] the locations of these points are distributed as order statistics from a sample of n is uniformly distributed the interval (0, t).

Theorem 2.6.3. If N is a homogeneous Poisson process in [0; +∞) with half , then conditioned N ((0, t]) = n, the number of points in the process in the interval [0, t] in increasing order is distributed as order statistics of a sample of size n is uniformly distributed in the interval [0, t], that is, the conditional distribution of the vector

$(S_1,\dots,S_n)$

As N ([0; T]) = n is the

same as the law

$$(U_{(1)},\dots,U_{(n)})$$

where the latter are of order statistics

variables

$$(U_{(1)},\dots,U_{(n)})$$

are uniformly distributed in [0; t].

Proof. Suppose that $a_1 < b_1 < a_2 < b_2 < \dots < a_n < b_n < t$,, then

$$P(S_i \in (a_i, b_i], i = 1, \dots n \mid N((0,t]) = n =$$

$$\frac{P(a_i < S_i < b_i, \forall i = 1, \dots, n, N((0,t]) = n)}{P(N((0,t]) = n)}$$

Now, focusing only on the numerator of 2.30 and applying the proposal 2.4.4 we have said numerator is:

$$P(N((0,a_1]) = 0, N((a_1,b_1]) = 1, \bigcap_{i=1}^{n-1} N((b_i,a_{i+1}]) = 0, N((a_{i+1},b_{i+1}]) = 1], N((b_n,t]) = 0)$$

$$= e^{-\lambda a_1} e^{-\lambda(b_1-a_1)} \lambda(b_1 - a_1)[\prod_{i=1}^{n-1} e^{-\lambda(a_{i+1}-b_i)} e^{-\lambda(a_{i+1}-b_{i+1})} \lambda(b_{i+1} - a_{i+1})] e^{-\lambda(t-b_n)}$$

$$= e^{-\lambda t} \lambda^n \prod_{i=1}^{n} (b_i - a_i)$$

2.10 then becomes:

$$e^{-\lambda t} \lambda^n \prod_{i=1}^{n} (b_i - a_i) \frac{n!}{e^{-\lambda t}(\lambda t)^n} = \frac{n!}{t^n} \prod_{i=1}^{n} (b_i - a_i)$$

Dividing the term

$$\prod_{i=1}^{n} (b_i - a_i)$$

$$b_i \downarrow a_i \forall i = 1, \dots n$$

given N ((0, t]) = n, then:

$$f_{S_1,\dots,S_n \mid N((0,t])}(a_1, \dots, a_n) = \frac{n!}{t^n}$$

by Lemma 2.6.2 that is what we wanted to try.

2.7. ALTERNATIVE DEFINITIONS OF POISSON PROCESSES

At the beginning of this chapter we saw a classic definition of a Poisson process. Immediately after we gave a constructive definition. In this section we compare the classical definition constructive.

Consider point processes defined in the whole real line R. We will note N (t) for N ((0, t]).

Definition 2.7.1. 1. We say that a one-dimensional process N (¢) has stationary if the distribution of the variable N ([s, t]) is the same as the variable

$N([s + r, t + r])\forall r \in R$

We say that the process N(¢) has independent increments if for

$A, B \subset R$

$A \cap B = \emptyset$

verified that the variables N (A) and N (B) are independent.

Definition 2.7.2. Let

$f: R \to R$

be a function. We say that the function f is one or (h) if:

$$\lim_{h\to 0} \frac{f(h)}{h} = 0$$

We will give below two definitions Poisson process and prove that they are equivalent.

Definition 2.7.3. We say that a process

$N(\cdot): \Omega \times P(R) \to N$

is a process *Poisson parameter ¸ if:*

1. (0)=0
2. The process has stationary and independent increments.
3. Random variables N([S, t]): *Ω×P(R)→N* have Poisson distribution with parameter (t s).

Definition 2.7.4. We say that a process (·):*Ω×P(R)→N* $N(\cdot): \Omega \times P(R) \to N$ is a process *Poisson parameter ¸ if:*

1. *N (0) = 0*
2. *The process has stationary and independent increments.*
3. *(N([t,t+h])=1)=λh+σ(h)* $P(N([t, t + h]) = 1) = \lambda h + \sigma(h)$
4. *(N([t,t+h])≥2)=σ(h)* $P(N([t, t + h]) \geq 2) = \sigma(h)$

Observation 2.7.5. Each of the above definitions univocally defines a process as a result of Theorem 2.3.1 (and is being said is that the distribution of the interval [s, t]).

Proposition 2.7.6. Definitions 2.7.3 and 2.7.4 are equivalent.

Proof. Let us first that involves 2.7.3 2.7.4.

We know that the variable N ([t, t + h]) has Poisson distribution with parameter H because the proposition 2.4.3. Then:

$$P(N([t, t+h])$$

$$P(N([t, t+h]) = 1) = \lambda h e^{-\lambda h} = \lambda h(1 + \sum_{i=1}^{\infty} \frac{(-\lambda h)^i}{i!})$$

using the series expansion

$e^{-\lambda h}$

Then if we find f (h) = that:

$$f(h) = \lambda h \sum_{i=1}^{\infty} \frac{(-\lambda h)^i}{i!}$$

and therefore f is a or (h) concluding that:

$$\lim_{h \to 0} \frac{f(h)}{h} = 0$$

$$P(N([t, t+h]) = 1) = \lambda h + o(h)$$

By similar reasoning we have:

$$P(N([t, t+h]) \geq 2) = \sum_{k=2}^{\infty} P(N([t, t+h] = k)$$

$$= \sum_{k=2}^{\infty} e^{-\lambda h} \frac{(\lambda h)^k}{k!}$$

$$= e^{-\lambda h} h^2 \sum_{k=2}^{\infty} \frac{\lambda^k h^{k-2}}{k!}$$

$$\lim_{h \to 0} \frac{P(N([t, t+h]) \geq 2)}{h} = \lim_{h \to 0} e^{-\lambda t} h \sum_{k=2}^{\infty} \frac{\lambda^k h^k}{k!} = 0$$

and therefore

$$P(N([t, t+h]) \geq 2) = o(h)$$

It remains to be seen now that involves 2.7.4 2.7.3. Note N (t) the total number of events occurred until time t is Pn (t) = P (N (t) = n). We can deduce a differential equation for P0 (t) as follows:

$$P_0(t+h) = P(N(t+h) = 0) = P(N(t) = 0, N(t+h) - N(t) = 0)$$

$$= P(N(t) = 0)\, P(N(t+h) - N(t) = 0)$$

$$= P_0(t)[1 - \lambda h + o(h)]$$

where the two last equalities are due to the conditions (3) and (4) of the definition 2.7.4. Then:

$$\lim_{h\to 0}\frac{P_0(t+h)-P_0(t)}{h}=\lim_{h\to 0}\frac{-\lambda hP_0(t)-P_0(t)o(h)}{h}$$

$$=\lim_{h\to 0}(\lambda P_0(t)+P_0(t)\frac{o(h)}{h})$$

$$=-\lambda P_0(t)$$

Therefore we concluded that

$$P`_0(t)=-\lambda P_0(t)$$

which is equivalent to:

$$\frac{P`_0(t)}{P_0\ (t)}=-\lambda$$

Integrating we obtain:

$$ln(\ P_0(t))\ =\ -\lambda t\ +\ c$$

for some constant c in R and then

$$P_0(t)\ =\ e^{-\lambda t}$$

where $k \in R$ $P_0(0)= P\ (\ N(0)= 0)= 1$then we have

$$P_0(t)\ =\ e^{-\lambda t}$$

Similarly, for $n\geq 1$ we have that:

$$P_n(t\ +\ h)\ =\ P(N(t\ +\ h)\ =\ n\ )$$

$$=\ P(N(t)=\ n,\qquad N(t\ +\ h)-\ N(t)\ =\ 0)$$

$$+\ P(N(t)=\ n-\ 1,\qquad N(t\ +\ h)-\ N(t)\ =\ 1)$$

$$+\ P(N(t\ +\ h)=\ n,\qquad N(t\ +\ h)-N(t)\ \geq\ 2$$

Now,

$$P(N(t\ +\ h)=\ n,\qquad N(t\ +\ h)-\ N(t)\ \geq\ 2)\ =\ P(N(t\ +\ h)\ =\ n,$$
$$N(h)\ \geq 2)\ \leq\ P(N(h)\ \geq\ 2)$$

Then, applying the condition (4) of the definition 2.7.4 we have:

$$\lim_{h\to 0}\frac{P(N(t+h)=n,N(t+h)-N(t)\geq 2}{h}\leq$$

$$\lim_{h\to 0}\frac{P(N(h)\geq 2)}{h}=0$$

then we conclude that $P(N(t+h)=n,N(t+h)-N(t)\geq 2)$

Then using the independence of the process increases we obtain

$$\lim_{h\to 0}\frac{P_n(t+h)-P_n(t)}{h}=\lim_{h\to 0}(-\lambda P_n(t)+\lambda P_{n-1}(t)+\frac{\hat{o}(\mathrm{h})}{h}$$

$$=-\lambda P_n(t)+\lambda P_{n-1}(t)$$

Therefore,

$$P`_n(t)=-\lambda P_n(t)+\lambda P_{n-1}(t)$$

Next we have

$$e^{\lambda t}(P`_n(t)+\lambda P_n(t))=e^{\lambda t}\lambda P_{n-1}(t)$$

$$\frac{d}{dt}(e^{\lambda t}P_n(t))=e^{\lambda t}\lambda P_{n-1}(t)$$

$$\frac{d}{dt}(e^{\lambda t}P_1(t))=e^{\lambda t}\lambda P_0(t)=\lambda$$

$$P_1(t)=(\lambda t+\hat{c})e^{\lambda t}$$

$$\hat{c}\in R$$

$$P_1(0)=P(N(0)=1)=0$$

$$P_1(t)=\lambda te^{-\lambda t}$$

$$P_n(t)=\frac{(\lambda t)^n}{n!}e^{-\lambda t}\ ,\forall n\in N$$

$$\frac{d}{dt}(e^{\lambda t}P_n(t))=e^{\lambda t}\lambda P_{n-1}(t)$$

$$\frac{d}{dt}(e^{\lambda t}P_n(t))=\lambda\frac{(\lambda t)^{n-1}}{(n-1)!}$$

$$e^{\lambda t}P_n(t)=\frac{(\lambda t)^n}{n!}+\tilde{c}$$

$$P_n(0)=P(N(0)=n)=0$$

$$P_n(t)=\frac{(\lambda t)^n}{n!}e^{-\lambda t}$$

As we wanted to see. It follows that the number of events at any interval of length t has Poisson distribution with parameter t which completes the test.

Theorem 2.7.7. The process

$$N(\cdot):\Omega\times P(R)\to N$$

defined in 2.2 is a Poisson process.

Proof. Proposition 2.4.3 know that the distribution of points in the interval [s; t] is Poisson parameter and the proposition 2.4.4 know that the process has stationary increments independent and therefore the process checks the properties of the definition 2.7.3.

2.8. POISSON PROCESSES IN TWO OR MORE DIMENSIONS

In this section we will build a Poisson process in two dimensions. This construction will be useful in the following chapters to build Poisson processes one-dimensional and overlapping of inhomogeneous Poisson processes.

We build analogously to what was done in one dimension, a random set of points in R2. The same construction that will below serves to build processes Rd with d ¸ 2.

Recall the construction made in the first sections of this chapter, considering a partition of R^2 A finite rectangles (For example, A_i could be

$$[n, n+1] \times [m, m+1] \qquad n, m \in Z$$

We denote by $|A_i|$ the area of the A-Rectangle. Each I in Z we assign a random variable

$$Y_i \colon \Omega \to N$$

Poisson distribution with $\lambda|A_i|$ for certain $\lambda \in R^+$we assume that these are independent variables. Then, for each i 2 Z we consider a succession of

$$\{U_{i,j}\}_{j \geq 1}$$

evenly distributed in the rectangle A_i. The assembly defined

$$S := \bigcup_{i \in Z} \left(\bigcup_{j=1}^{Y_i} \{U_{i,j}\} \right)$$

$$\bigcup_{j=1}^{0} \{U_{i,j}\} = \emptyset$$

In other words we do is placed in each A_i an amount of points evenly distributed. Until now

We repeat the procedure made for one dimension. The difference now is that there is no satisfactory way to sort the points in R^2, But that is not important. Now let A ½ R^2a measurable set. We define the two-dimensional Poisson process:

$$M: \Omega \times P(R^2) \to N$$

such that: $M(w, A) = M_{S(w)}(A) :=$ 1 number of points in the set $S(w) \cap A$

Then

$$M(A): \Omega \to N$$

counts the number of points in the set S \ A. To avoid confusion we will use the letter M dimensional Poisson processes and letter for one-dimensional.

The following properties were tested with the same ideas used for one-dimensional case, therefore omit their proofs.

Proposition 2.8.1. For each finite set A ½ R2 measurable random variable M (A): Ω! N has Poisson distribution of parameter JAJ.

Proposition 2.8.2. For each finite family of disjoint measurable sets B_1; … B_L *in* R2 random variables $M(B_l): \Omega \to N$

with

$$l = 1, \dots L$$

have Poisson distribution with $\lambda|B_l|$ and are independent.

Proposition 2.8.3. For any measurable set A ⊂ R2 conditional distribution points S \ A given M (A) = n is the same as that of n uniformly distributed vai set A.

Example 2.8.4. Given the construction in 2 dimensions we can calculate the law of the random variable

$$V: \Omega \to \mathbb{R}$$

defined by

$$V(w) = \inf\{|x| : x \in S(w)\}$$

measuring the distance from point p that is closest to the origin.

Be

$$B(0, b) := \left\{(x, y) \in \mathbb{R}^2 : \sqrt{x^2 + y^2} < b\right\}$$

ball center the origin and radius b in R+.

So:

$P(V > b) = P(\{w \in \Omega: V(w) > b\})$

Now look to ask V> b is asking the infimum of the distances of the points of *S*(w) The origin is greater than b in R +, which is equivalent to requiring that B (0, b) no points S (w). Then 2.36 is equivalent to:

$P\ (M\ (B\ (0, b))\ =\ 0)$

But we know that the variable $M(B\ (0, b)): \Omega -> N$ has Poisson distribution with parameter ,

$\lambda\left(area(B(0, b))\right) = \lambda b^2 \pi$

$P(V > b) = e^{-\lambda \pi b^2}$

2.9. PROJECTIONS

In this section we show a way of building Poisson processes one-dimensional from two-dimensional processes.

Let

$M(\cdot): \Omega \times P(R) \to N$

be N dimensional Poisson process with parameter , = 1.

We want to build a process of dimension one $N(I) := M(I \times [0, \lambda])$with parameter

$I \times [0, \lambda]$

as a function of M (¢). We define for each interval I ½ R:

$N(\cdot): \Omega \times P(R) \to N$

such that N (I) $= M(I \times [0, \lambda])$

Then the number of points of N (¢) process in the interval I is the number of points M (¢) process in the rectangle.

The following proposition shows that the process defined by 2.37 is indeed a Poisson process one dimension.

Proposition 2.9.1. The process $N(\cdot): \Omega \times P(R) \to N$is defined by a Poisson process one-dimensional parameter

Proof. Observe first that the random variable N ([a, b]): Ω! N has Poisson distribution with parameter

$\lambda(b-a)$

by Proposition 2.8.1. Consider secondly the process has independent increments. Let

$I, J \subset R$

$I \cap J = \emptyset$

Then as the process M (¢) is Poisson (therefore has independent and

$(I \times [0,\lambda] \cap (J \times [0,\lambda]) = \emptyset$

we have:

$$P(N(I) = k, N(J) = h) = P(M(I \times [0,\lambda]) = k), M(J \times [0,\lambda]) = h)$$
$$= P(M(I \times [0,\lambda]) = k)P(M(J \times [0,\lambda]) = h)$$
$$= P(N(I) = k)P(N(J) = h)$$

which means that has independent increases.

Finally we see that has stationary increments. Be [a; b] $\subset$R interval k in N, then:

$$P(N([a,b]) = k) = P(M([a,b] \times [0,\lambda]) = k) = \frac{(\lambda(b-a))^k}{k!}e^{-\lambda(b-a)} \qquad (2.7)$$

where the last equality is due to proposition 2.8.1.

On the other hand, since r in R, making a similar argument we have:

$$P(N([a+r,b+r]) = k) = P(M([a+r,b+r] \times [0,\lambda]) = k) = \frac{(\lambda(b-a))^k}{k!}e^{-\lambda(b-a)} \qquad (2.8)$$

Then as 2.7 and 2.8 are equal, the process has stationary increments. Thus verified for N (¢) process the definition 2.7.3.ı

A question that may arise is if two points of a two-dimensional process may be projected at the same point of a one-dimensional process. The answer to this question is given in the following lemma.

Lemma 2.9.2. I $\subset$R is a finite interval. The event "two points of a process-Poisson are two dimensional are projected onto the same point interval I" has probability-0.

Proof. Suppose without loss of generality that I = [0; 1]. we partition I

$$\delta < 1: I_n^\delta := ((n-1)\delta, n\delta]$$

$$\{I_n^\delta\}_{n=1}^{n=\frac{|I|}{\delta}}$$

Consider P (Two points are projected at the same point)

P (Two points belong to the same interval

$$\{I_n^\delta\}_{n=1}^{n=\frac{|I|}{\delta}}$$

are equal to

$$P(\bigcup_{n=1}^{\frac{|I|}{\delta}}\{M(I_n^\delta \times [0,\lambda]) \geq 2\})$$

Applying the proposition 2.8.1 and using the area

$$I_n^\delta \times [0,\lambda]$$

it is , ± for all n we have to 2.40 this bounded above by:

$$\sum_{n=1}^{\frac{|I|}{\delta}}(\sum_{j=2}^{+\infty}\frac{(\lambda\delta)^j}{j!}e^{-\lambda\delta}) = [\frac{|I|}{\delta}]e^{-\lambda\delta}\sum_{j=2}^{+\infty}\frac{(\lambda\delta)^j}{j!}$$

Let now f: R -> R such that f (±) = e. It then holds that

$$f(\delta) = e^{-\lambda\delta}\sum_{j=2}^{+\infty}\frac{(\lambda\delta)^j}{j!}$$

$$\lim_{\delta \to 0}\frac{f(\delta)}{\delta} = 0$$

and therefore f is a $o(\delta)$Then we have:

$$P \leq) = [\frac{|I|}{\delta}]o(\delta) \to 0$$

P (Two points are projected at the same point)

$$P \leq) = [\frac{|I|}{\delta}]o(\delta) \rightarrow 0$$

when $\delta \rightarrow 0$ which completes the proof.

Observation 2.9.3. The construction of one-dimensional processes from bidi-dimensional has the advantage that allows us to simultaneously build processes with different parameters, so that one has always greater or equal number of points than the other, as seen in the following lemma.

Lemma 2.9.4. Let M (¢) two-dimensional Poisson process, and are 1; 2 2 R + 1 such that , 2. Poisson processes define Ni with i = 1; 2 such that:

$$N_i(I) = M(I \times [0, \lambda_i])$$

to I = 1; 2 and for all I subset of R. Then it holds that:

$$N_1(I) \geq N_2(I)\ \forall I \subset R$$

Proof. By definition it is clear that N1 2.37 greater than or equal projected amount of points N2.

2.10. POISSON PROCESS

Suppose a bank male clients arrive according to a Poisson process of parameter PY female clients according to a Poisson process of parameter , (1 Hot p) for some 0 <p <1. One way to model the process arrivals, according sex distinguishing arrivals process is to construct a two-dimensional Poisson M (¢) and define processes $N_{1\,(¢)}$ and N_2(¢) as follows:

$$N_1(I)\colon \Omega \rightarrow N$$

$$N_1(I) = M(I \times [0, \lambda_p])$$

$$N_2(I)\colon \Omega \rightarrow N$$

$$N_2(I) = M(I \times [\lambda_p, \lambda])$$

for all I subset of R. In this way the points are in the strip $R \times [0, \lambda]$ are projected and indicate the instants of arrival of the customers, regardless of gender of each points coming from the strip

$$R \times [0, \lambda_p]$$

mark them and determine.

Arrival times men customers, and points coming from the $R \times [\lambda_p, \lambda]$ the mark with 0 and indicate the times of arrival of women customers. The process N (¢) such that

$$N(I) := N_1(I) + N_2(I)$$

it can be written as

$$N(I) = M(I \times [0, \lambda])$$

therefore is a process Poisson parameter.

Observe Sn with n >= 1 arrival times of customers regardless of gender, i.e., S1 is the arrival time of the first customer to the bank and Sn is the time of arrival to the Nth bank customer. Each of the above has a brand that is 0 or 1 depending on whether we are male or female respectively. Be the variables

$$G(S_i): \Omega \to \{0,1\}$$

such that

$$G(S_i) := \begin{cases} 1 \\ 0 \end{cases}$$

Proposition 2.10.1. Random variables $G(S_i): \Omega \to \{0,1\}$ are with law:

$$P(G(S_i) = 1) = p$$

$$P(G(S_i) = 0) = 1 - p$$

for all i >= 1.

Proof. Build to start n valid uniform in the rectangle [0; t] x [0; t]. FVngn¸1 and fWngn¸1 are two independent sequences will be uniformly distributions in the range [0; 1] and independent each other. For each n, is

$$\pi_n: \{1,2, \dots, n\} \to \{1,2, \dots, n\}$$

a bijective function (i.e. Pin is a permutation of n elements) defined by: for $\{1,2, \dots, n\}$

holds that

$$V_{\pi_{n(i)}} \leq V_{\pi_{n(i+1)}}$$

In other words, if increasingly ordain the first n variables

$$V_1, \dots, V_n$$

$\pi_n(i)$ is the subscript of the i-th variable.

For each fixed n, construct a succession of runs [0; t] x [0; t] as follows:

We define $Ui\colon \Omega! -> \ [0;\ t]\ x\ [0, t\ ]$ such that

$$U_i := \left(tV_{\pi_{n(i)}}, \lambda W_i\right) \qquad \forall i = 1, \dots, n$$

Let the variables

$$\{U_i\}i=1,\dots n$$

are independent and uniformly distributed in the rectangle-mind [0; t] x [0; t]. As the variables Vi and Wj are independent variables, we have:

$$P(U_i \in [a,b] \times [c,d]) = P(tV_{\pi_{n(i)}} \in [a,b], \lambda W_i \in [c,d])$$

$$= P(tV_{\pi_{n(i)}} \in [a,b]P(\lambda W_i \in [c,d])$$

$$= P\left(\frac{a}{t} \leq V_{\pi_{n(i)}} \leq \frac{b}{t}\right) P(\frac{c}{t} \leq W_i \leq \frac{d}{\lambda})$$

Then, as we know the distribution of variables V_i and W_j. It is easy to see that:

$$U_i \sim ([0,t] \times [0,\lambda]) \, \forall i = 1, \dots n$$

It remains to be seen to be independent, which follows easily from the fact that

$$\{V_i\}_{i \geq 1}$$

$$\{W_j\}_{j \geq 1}$$

are families who are also independent. Then build $\{U_i\}i=1,\dots n$ a family of n valid uniform [0; t] x [0; t]. Note that if the variables ordered n increasingly according to the abscissa

$$(V_{\pi_{n(i)}})$$

then the ordered Wi is the i-th of Uj. It is now fixed N and

$$L \in N$$

$$1 \leq k \leq L$$

$$a_k \in \{0,1\}$$

arbitrary. We define the set

$$A_L := \{G(S_1) = a_1, \dots, G(S_L) = a_L\}$$

Using a discrete version of Theorem 2.3.1, to prove that G (Si) with variables i ¸ 1 have Bernoulli distribution of parameter p reaches to prove that:

$$P(A_L) = p^{P_{a_k}}(1-p)^{P_{(1-a_k)}}$$

Let t> 0 fixed. So

$$P(A_L) = \sum_{n=L}^{+\infty} P\big(A_L, N\big((0,t)\big) = n\big) + P(A_L, N((0,t)) < L)$$

Now

$$P\big(A_L, N\big((0,t)\big) = n\big) = P\big(A_L \mid N\big((0,t)\big) = n\big)P\big(N\big((0,t)\big) = n\big)$$

By definition

$$N((0,t)) = M([0,t] \times [0,\lambda] = n$$

and the proposition 2.8.3 distribution of points in this rectangle as M ([0; T] x [0; T]) = n it is the same as the distribution of n valid uniform in the rectangle. For 1 · k · L are

$$I_k = \begin{cases} [0,p\lambda] \ \ if \ \ a_k = 1 \\ [p\lambda,\lambda] \ \ if \ a_k = 0 \end{cases}$$

Then, using the construction 2.44 and Proposition 2.8.3 we have:

$$\begin{aligned} P\big(A_L \mid N\big((0,t)\big) = n\big) &= P(A_L \mid M([0,t] \times I_k, 1 \leq k \leq n) \\ &= P(U_{\pi_n(k)} \in [0,t] \times I_k, 1 \leq k \leq n) \\ &= P(\lambda W_k \in I_k, 1 \leq k \leq n) \\ &= p^{P_{a_k}}(1-p)^{P_{(1-a_k)}} \end{aligned}$$

where the last equality is because:

$$P(\lambda W_k \in I_k) = \frac{|I_k|}{\lambda} = \begin{cases} p \quad\quad\ if \ \ a_k = 1 \\ 1-p \ \ \ if \ a_k = 0 \end{cases}$$

then as this equality does not depend on n we have:

$$P(A_L) = \sum_{n=L}^{+\infty} p^{P_{a_K}}(1-p)^{P_{(1-a_K)}} P\big(N((0,t)) = n\big) + P(A_L, N((0,t)) < L)$$

$$= p^{P_{a_K}}(1-p)^{P_{(1-a_K)}} \sum_{n=L}^{+\infty} P\big(N((0,t)) = n\big) + P(A_L, N((0,t)) < L)$$

$$= p^{P_{a_K}}(1-p)^{P_{(1-a_K)}} P\big(N((0,t)) \geq L\big) + P(A_L, N((0,t)) < L)$$

By the law of large numbers for a Poisson process one-dimensional [1] is true:

$$\frac{N((0,t))}{t} \to \lambda$$

$$\lim_{t\to\infty} P\big(N((0,t)) \geq L\big) = 1$$

$$\lim_{t\to\infty} P\big(A_L, N((0,t)) < L\big) = 0$$

Then

$$P(A_L) = p^{P_{a_K}}(1-p)^{P_{(1-a_K)}}$$

As we wanted to see. We conclude that G (Si) variables; $P(A_L)$are of Bernoulli distribution of parameter p.

Analogously to what was done in the section (2.5), let us now see one alternative construction of a Poisson process two-dimensional M (¢) with parameter 1 in the belt [0; 1] x [0; 1]. This construction will be used in the proof of the following theorem.

T1 will be one parameter exponential W1 is uniform one in the range [0; 1] and M1 (¢) a Poisson process with two-dimensional parameter defined in the belt

$[0, \infty] \times [0, \lambda]$

We assume independence between T1; W1 and M1 (¢). We define the M process (¢) as:

$$M(A) := 1_{\{(T_i, W_i) \in A\}} + M_1(A - T_1)$$

$$A \subset R^2$$

$$A - t = \{(x,y) \in R^2 : (x+t, y) \in A\}$$

arguments similar to those used in Theorem 2.5.1, 2.5.2 corollary, and proposition 2.10.1 prove the following theorem.

Theorem 2.10.2. Let M (¢) two-dimensional Poisson process with parameter 1. Observe *ordered times of occurrence of events in the strip* [0; ,] and are {W,} *the second coordinates of these events, then*

$\{S_{i+1} - S_i\}_{i \geq 1}$

are valid distributions exponentially with parameter are valid with uniform distribution [0;].

2.11. PROCESSES WHICH ARE NOT HOMOGENOUS

Be a non-negative piecewise differentiable function. Suppose for each finite interval I ½ , R function has a finite number of discontinuities in I. We construct a point process with independent increments and "instantaneous rate" , (t), this means that we want a process N (¢): Ω £ P (R)! N such that at points of continuity of the function , check that:

$$P(N([t, t+1]) = 1) = h\lambda(t) + o(h)$$

$$P(N([t, t+1]) \geq 2) = o(h)$$

For this, consider a two-dimensional process and define

$$M(\cdot): \Omega \times P(R^2) \to N$$

one-dimensional process $N(\cdot): \Omega \times P(R) \to N$ such that

$$N(I) := M(F(I)) \quad \forall I \subset R$$

where

$$F(I) := \{(x, y) \in R^2 : x \in I, 0 \leq y \leq \lambda(x)\}$$

Thus N (¢) is the process of projecting the points M (¢) which are "below" , function.

Lemma 2.11.1. The process verifies the conditions 2.49 2.47 and 2.48

Proof. Let t from R a point of continuity of the function. By the definition of *N*(¢) is satisfied that:

$$P(N([t, t+1]) = 1) = P(M(F([t, t+h]\}) = 1)$$

Observe

$$y_0 := y_0(t, h) = \inf\{\lambda(x): x \in [t, t+h]\}$$

$$y_1 := y_1(t, h) = sup\{\lambda(x): x \in [t, t+h]\}$$

Then it is clear that:

$$M([t, t+1] \times [0, y_0]) \leq M(F([t, t+h])) \leq M([t, t+h] \times [0, y_1])$$

On the other hand, as , function it is continuous in t we have:

$$\lim_{h \to 0} y_0(t, h) = \lambda(t)$$

$$\lim_{h \to 0} y_1(t, h) = \lambda(t)$$

And also by differentiability

$$y_1(t, h) - y_0(t, h) = O(h)$$

where O (h) indicates a function h such that lim*h!*0 OR(H). It is bounded above and below. We then have:

$$\lim_{h \to 0} \frac{O(h)}{h}$$

$$P(M([t, t+h] \times [0, y_1]) - M([t, t+h] \times [0, y_0]) > 1 = o(h)$$

In this way

$$P(M(F([t, t+h])) = 1, M([t, t+h] \times [0, y_1]) - M([t, t+h] \times [0, y_0]) = 0)$$
$$+ P(M(F([t, t+h])) = 1, M([t, t+h] \times [0, y_1]) - M([t, t+h] \times [0, y_0]) \geq 1)$$

(2.9)

The first term of 2.9 equals:

$$P(M([t, t+h]) \times [0, y_0]) = 1, M([t, t+h] \times [0, y_1]) - M([t, t+h] \times [0, y_0]) = 0)$$

and as the process M (¢) has independent increments that gives us:

$$P(M([t, t+h]) \times [0, y_0]) = 1, P(M([t, t+h] \times [0, y_1]) - M([t, t+h] \times [0, y_0]) = 0) =$$
$$hy_o(t, h)^{-hy_0(t,h)}(1 - o(h)) =$$
$$hy_o(t, h)^{-hy_0(t,h)} - hy_o(t, h)^{-hy_0(t,h)}o(h)$$
$$P(M([t, t+h]) \times [0, y_1]) - M([t, t+h] \times [0, y_0]) \geq 1) = o(h) + h(y_1 - y_0)$$

The second term is bounded by:

$$P(M(F(t, t+h) = 1)) = hy_o(t, h)^{-hy_0(t,h)} - hy_o(t, h)^{-hy_0(t,h)}o(h) + h(y_1 - y_0) = h\lambda(t) + o(h)$$

Now notice that the Poisson process constructed above has the property that the number of events in any interval is a variable parameter Poisson equal to the area under the function , (t) in the interval, i.e.,

$$P(N((0, t]) = n) = P(M(F([0, t])) = n)$$

and we know that

$$M(F([0,t]))$$

is a Poisson variable with parameter the area of the region F ([0; T]) which is nothing but R0t , (s) ds.

Analogous to the provisions of 2.9.3 observation, the definition 2.49 has the advantage that allows us to jointly build two or more processes with different rates and the following property:

Proposition 2.11.2. Observe 1 (t) and 2 (t) two continuous piecewise functions that satisfy the condition

$$\mu(t) = \int_0^t \lambda(s)ds$$

Then it is possible to jointly build inhomogeneous Poisson process

$$P(N((0,t]) = n) = P(M(F([0,t]) = n)$$

rate 1 (t) and 2 (t) respectively such that for each interval I ½ R is satisfied that

N_1(I) · N2 (I)

Proof. It follows directly from the definition of inhomogeneous process.

CHAPTER 3

MARKOV CHAINS IN CONTINUOUS TIME

CONTENTS

In this chapter we will consider Markov chains in continuous time. As in the case of discrete-time chains, these are characterized by the Markovian property that given the present state, the future is independent of the past.

The first sections define Markov chains continuous time and see how to build a-e have those properties process. The fourth section stable two sets of differential equations -backward and forward equations- describing the law of probability of the system. The rest of the chapter is intended to show results and interesting examples of theory.

3.1. PURE MARKOV PROCESSES

Definition 3.1.1. Observe a stochastic process $\{X_t\}_{t\geq 0}$ continuous time (i.e., t in [0; T] with fixed T in R), which takes values in a countable set E. We say that the process $\{X_t)$ is a Markov chain if continuous time

$0 \leq u \leq s$

is satisfied that

$$P(X_{t+s=j}|X_s = i, X_u = x_u \ \forall 0 \leq u \leq s) = P(X_{t+s} = j|X_s = i)$$

In other words, a string of continuous-time Markov (CMTC) is a stochastic process that verifies the Markovian property, that is, the conditional probability of a future state at time t + s, given the present state at time s all past states, only depends on the present state and independent of the past.

Definition 3.1.2. If

$$\{X_t\}_{t\geq 0}$$

$$0 \leq u \leq s$$

is a CMTC and also verifies that

$$P(X_{t+s} = j|X_s = i)$$

is independent of s then said that the CMTC has stationary transition probabilities or homogeneous.

We then construct a process $\{X_t\}$ values in a space E numerable and then see that this construction makes and suppose then that the state space E is a countable set. Let q

$q: E \times E \to R$

$q(x,y) \geq 0$

$x, y \in E$

$x \neq y$

call (x, y) transition speed state and the state x. We want to build a process with the property that the speed jump from state q other and is (x, y). In other words the process should check

$$P(X_{t+h} = y | X_t = x) = hq(x,y) + o(h) \forall x \neq y$$

Let

$$q: E \to R$$

$$q(x) := \sum_{y \in E} q(x,y)$$

and q (x, y) the output speed of the state x. Assume that output rates are uniformly bounded, i.e.,

$$\lambda := \sup_{x \in E} \sum_{y \in E} q(x,y) = \sup_{x \in E} q(x) < +\infty$$

(Note that if the state space E is finite, then this condition is automatically satisfied). To build the process, we introduce a Poisson process bidi-dimensional

$$M(\cdot): \Omega \times P(R^2) \to R$$

with parameter 1. For each x in E we partition the interval

$$I^x := [0, q(x)]$$

at intervals I (x, y) such that

$$|I(x,y)| = q(x,y)$$

Thus, we have

$$\{I(x,y)\}_{y \in E} \subset R$$

verifies that:

$1. I(x,y)| = q(x,y)$

$2. \cup_{y \in E} I(x,y) = I^x = [0, q(x)]$

Be x_0 from E and assume that $X_0 = x_0$. Sea τ_1 the first time an event of M process

$$I^{x_0}$$

appears in the Ix_0 interval. So:

$\inf\{t > 0: M([0,t] \times I^{x_0} > 0\}$

X_1 = y define where and is the only state that satisfies

$$\inf\{t > 0: M([0,t] \times I^{x_0} > 0\} = \inf\{t > 0: M([0,t] \times I(x_0, y)) > 0\} \quad (3.6)$$

(Note that x1 is well defined, as

$\{I(x,y)\}_{y \in E}$

partition the interval [0, q (x)] and then verifying the state is only 3.6). Suppose now that we have certain

x_{n-1}

τ_{n-1}

It then define inductively:

$\tau_n := \inf\{t > \tau_{n-1}: M([\tau_{n-1}, t] \times I^{x_{n-1}}) > 0\}$

and x_n = y where y is the only state that verifies that

$\inf\{t > \tau_{n-1}: M([\tau_{n-1}, t] \times I^{x_{n-1}}) > 0\} = \inf\{t > \tau_{n-1}: M([\tau_{n-1}, t] \times I(x_{n-1}, y)) > 0\}$

Definition 3.1.3. We define

$\tau_\infty := \sup_n \tau_n$

$X_t: \Omega \to E$

$t \in [0, \tau_\infty)$

and the process

$X_T := x_n$

such that

$t \in [\tau_n, \tau_{n+1})$

Thus zn is the time when the Nth jump process occurs, and xn is the nth been visited by

$\{X_t\}_{t \in [0,\infty)}$

Example 3.1.4. Let's see how to do the above construction for a process with 3 states. Suppose we have a system with 2 servers, which has no place waiting for customers not served, i.e., customers who arrive when the two servers are busy leave the system automatically. Suppose also that customers arrive at the system according to a Poisson process of rate ¸ and servers serve an exponential distribution with parameter. It is clear that this system has 3 possible states: no customers in the system; there is a customer in the system; there are 2 customers in the system. Let then

$$E = \{1,2,3\}$$

our state space. Which we are now calculate speeds transition between states.

1. If the system is in state 0, i.e., no customers, there can be only transfer to state 1 when a customer arrives, which occurs ¸ rate.
2. *If there is a customer in the system, then the next state may be 2 clients (in case we get a new customer to the system, which occurs with ¸ rate), or it may be that the system returns to state 0 (in the event that the customer who is being treated is removed before it reaches the new, which occurs with rate ¯).*
3. *If there are 2 customers in the system then we can only go to 1 when a client finishes injury, which happens rate* 2 since there are two servers serving, each cup ¯.

So we have to

1. $q(0,1) = \lambda$
2. $q(1,2) = \lambda$
3. $q(1,0) = \lambda$
4. $q(2,1) = 2\beta$

and other transition rates are 0. Note that the speeds of the states are bounded

$$\max\{\lambda + \beta, 2\beta\}$$

The next step is to partition intervals [0; q (x)] for each

$$x \in E$$

intervals

I(X, y) of length q (x, y). Let one example, others are analogous. We know that

$$q(1) = \sum_{y=0}^{2} q(1,y) = q(1,0) + q(1,1) + q(1,2) = \beta + \lambda$$

Then, a possible choice of intervals could be the following:

$|I(1,2)| = q(1,2) = \lambda$

$I(1,2) = [0,\lambda]$

$|I(1,0)| = q(1,0) = \beta$

Moreover we know

$|I(1,1)| = 0$

$|I(1,0)| = 0$

then define

$I(1,0) = [\lambda, \lambda + \beta]$

$I(1,0) = \emptyset$

$I(0,1) = I(1,2) = [0,\lambda]$

Following this reasoning intervals can be chosen:

$I(1,0) = [\lambda, \lambda + \beta]$

$I(1,0) = \emptyset$

$I(0,1) = I(1,2) = [0,\lambda]$

$I(0,0) + I(0,2) = I(1,1) = I(2,0) + I(2,2) = \emptyset$

Then, according to a Poisson process M(¢) two intervals before the $\{X_t\}_{t\in[0,\tau_\infty)}$

Let's look at the process defined in 3.10 verifies the condition 3.3.

Proposition 3.1.5. The

$\{X_t\}_{t\in[0,\tau_\infty)}$

Is defined by 3.10 verifies ownership

$P(X_{t+h} = y | X_t = x) = hq(x,y) + o(h)\ \forall x \neq y$

Proof. By the definition of the process, it is clear that:

$$\{X_{t+h} = y | X_t = x\} \subset \{M((t,t+h] \times I(x,y)) = 1\} \bigcup \{M((t,t+h] \times [0,q(x)]) \geq 2\}$$

Then it holds that:

$$P(X_{t+h} = y|X_t = x) = P\big(M\big((t, t+h] \times I(x,y)\big) = 1\big) + P(A)$$
$$A \subset \{M((t, t+h] \times [0, q(x)]) \geq 2\} \tag{3.11}$$

Furthermore, as q (x) is bounded by applying the proposition 2.8.1 we have:

$$P\big(M\big((t, t+h] \times I(x,y)\big) = 1\big) = h|I(x,y)|e^{-h|I(x,y)|} =$$
$$= hq(x,y)e^{-hq(x,y)}$$
$$= hq(x,y) + o(h)$$

$$P\big(M\big((t, t+h] \times [0, q(x)]\big) \geq 2\big) = 1 - P\big(M\big((t, t+h] \times [0, q(x)]\big) = 0\big)$$
$$= 1 - e^{-hq(x)} - hq(x)e^{-hq(x)}$$
$$= o(h)$$

where the last equality in each equation is deduced by using the series expansion of the exponential function. Using the above result, ending only be noted that as

$$A \subset \{M((t, t+h] \times [0, q(x)]) \geq 2)$$
$$P(A) \leq P(M((t, t+h] \times [0, q(x)]) \geq 2)$$

Then

$$\lim_{h \to 0} \frac{P(A)}{h} \leq \lim_{h \to 0} \frac{P(M((t, t+h] \times [0, q(x)]) \geq 2)}{h} = 0$$

whereby P (A) is a or (h). Returning to 3.11 then we conclude that

$$P(X_{t+h} = y|X_t = x) = hg(x,y) + o(h)$$

for all x; are from E different, as we wanted to prove.

3.2 PROPERTIES

In this section we will see that the process Xt constructed by 3.10 is a Markov chain of continuous time, i.e., given the present, past and the future are independent.

Proposition 3.2.1. The {Xt [0 process; 1) Is defined condition 3.10 satisfies the following:

$$P\big(X_{t+s} = y|X_t = x, X_u = x_u\ \forall 0 \leq u \leq t\big) = P(X_{t+s} = y|\ X_t = x)$$

where $x, y, x_u \in E, \forall 0 \leq u \leq t$

Proof. For construction made, the process after the time t_0 only depends on the two dimensional process M (¢) in the region (t0; +1) £ R and the value taken X_t at time t_0. Then, given the assumed value chain at time t0, the future is independent of the past.

In the next two results we will see that:

Since the process X_t It is in the state x from E at the time, The elapsed time until the next hop is going exponential distribution average q.

$x, y, x_u \in E, \forall 0 \leq u \leq t$

$(t_0, +\infty)$

$$\frac{1}{q(x)}$$

when the process decides to jump a makes with probability

$$p(x, y) := \frac{q(x, y)}{q(x)}$$

Theorem 3.2.2. It is a process of continuous-time Markov such that the output speeds of states q (x) are uniformly bounded by ,> 0. Then better:

$$P(\tau_{n+1} - \tau_n > t | X_{\tau_n} = x) = e^{-tq(x)}$$

Proof. Let

$$M(\cdot): \Omega \times P(R^2) \to N$$

be N-dimensional Poisson process of parameter to one. Let Sn with n from N time ordered occurrence of even-cough in the range

$[0, \lambda]$

and n are Wn with 2N second coordinates of those events (as in Theorem 2.10.2). By Theorem 2.10.2 we know that

1.{ $\{\{S_{i+1} - S_i\}_{i \geq 1}$ they are distributed exponentially with parameter

2. $\{W\}_{i \geq 1}$ they are uniformly distributed in the interval [0; 1]

3. $\{\{S_{i+1} - S_i\}_{i \geq 1}$ $\{W\}_{i \geq 1}$are independent.

It is clear that the set

$$\{\tau_n\}_{n\geq 0} \subset \{S_n\}_{n\geq 0}$$

$$x_0 \in E$$

In fact, given x from E can define:

$$\tau_n := \min\{S_k > \tau_{n-1}: W_k < q(x_{n-1})\}$$

$$K_n := \{k: S_k = \tau_n\}$$

$$x_n := \{y \in E: W_{K_n} \in I(x_{n-1,}y)\}$$

The distribution

$$\tau_{n+1}$$

$$\tau_n$$

N conditioned on $X_{\tau_n} = x$ is

$$P(\tau_{n+1} - \tau_n > t, X_{\tau_n} = x) = \frac{P(\tau_{n+1} - \tau_n > t, X_{\tau_n} = x)}{P(X_{\tau_n} = x)}$$

Conditioning the possible values that Kn can take, we must

$$P(\tau_{n+1} - \tau_n > t, X_{\tau_n} = x) = \sum_{k\in N} P(\tau_{n+1} - \tau_n > t, X_{\tau_n} = x, K_n = k)$$

which, using the definition of Kn is the same as

$$\sum_{k\in N} P(\tau_{n+1} - S_k > t, X_{S_k} = x, K_n = k)$$

Using now the definition and results 3.14 (1), (2) and (3) about

$$\{S_{i+1} - S_i\}_{i\geq 1}$$

$$\{W_i\}_{i\geq 1}$$

variables and statements at the beginning of the show, we have the above is:

$$\sum_{k\in N}(\sum_{l\in N} P(\{S_{k+l} - S_k\} > t, W_{k+1} > q(x), \ldots, W_{k+l-1} > q(x), W_{k+l} < q(x)))\ \ P(X_{S_k} = x, K_n = k)$$

$$= e^{-tq(x)} \sum_{k\in N} P(X_{S_k} = x, K_n = k)$$

So we have that

$$P(\tau_{n+1} - \tau_n > t \mid X_{\tau_n} = x) = \frac{e^{-tq(x)}}{P(X_{\tau_n} = x)} \sum_{k \in N} P(X_{S_k} = x, K_n = k)$$

$$= \frac{e^{-tq(x)}}{P(X_{\tau_n} = x)} \sum_{k \in N} P(X_{\tau_n} = x, K_n = k)$$

$$= \frac{e^{-tq(x)}}{P(X_{\tau_n} = x)} \sum_{k \in N} P(X_{\tau_n} = x)$$

$$= e^{-tq(x)}$$

as we wanted to prove.

Definition 3.2.3. We define the discrete-time process Yn: Ω! N N n 2 so that Yn = X¿n 8 n 2 N. This process will be called skeleton fXtgt¸0 continuous time process.

Theorem 3.2.4. The $Y_n = X_{\tau_n} \ \forall n \in N$ skeleton of a continuous time process Xt is a Markov chain with transition probabilities fp (x, y): x; and 2 given by, e.g.,

p(X, y): =q(X, y)

Proof. We want to prove that

$$P(Y_{n+1} = y \mid Y_n = x) = p(x, y)$$

which is equivalent prove that

$$P(Y_{n+1} = y \mid Y_n = x) = p(x, y)$$

$$P(X_{\tau_{n+1}} = y \mid X_{\tau_n} = x) = p(x, y)$$

From 3.14, 3.15 and 3.16. Starting the space according to the values that can take K_n we have

$$P(X_{\tau_{n+1}} = y \mid X_{\tau_n} = x) = \sum_{k \in N} P(X_{\tau_{n+1}} = y, X_{\tau_n} = x, K_n = k)$$

$$\{X_{\tau_{n+1}} = y, X_{\tau_n} = x, K_n = k\}$$

Now notice that as Kn = k then meets the set fX¿n + 1 = y; X¿n = x; Kn = kg is

$$\bigcup_{l\geq 1}\{W_{k+1} > q(x), \dots, W_{k+l-1} > q(x),\ W_{k+l} \in I(x,y), X_{S_k} = x, K_n = k\}$$

$$\sum_{k\in N}\sum_{l\geq 1} P(W_{k+l-1} > q(x), W_{k+l} \in I(x,y), X_{S_k} = x, K_n = k\}$$

Using the above along with the independence between fWigi2N and 3.18 fSigi2N we have is

$$\sum_{l\geq 1} P\big(W_{k+1} > q(x), \dots, W_{k+l-1} > q(x), W_{k+l} \in I(x,y)\big) =$$

$$\sum_{l\geq 1}(\frac{\lambda - q(x)}{\lambda})^{l-1}\ \frac{|I(x,y)|}{\lambda} = \frac{q(x,y)}{q(x)}$$

As we have

$\{W_i\}_{i\geq 1}$

are independent

$$P\big(Y_{n+1} = y \mid Y_n = x\big) = p(x,y)$$

$$P\big(Y_{n+1} = y \mid Y_n = x\big) = p(x,y)$$

$$P\big(X_{\tau_{n+1}} = y \mid X_{\tau_n} = x\big) = p(x,y)$$

$$P\big(X_{\tau_{n+1}} = y \mid X_{\tau_n} = x\big) = \sum_{k\in N} P(X_{\tau_{n+1}} = y\ , X_{\tau_n} = x, K_n = k)$$

$$\{X_{\tau_{n+1}} = y\ , X_{\tau_n} = x, K_n = k\}$$

$$\sum_{l\geq 1} P\big(W_{k+1} > q(x), \dots, W_{k+l-1} > q(x), W_{k+l} \in I(x,y)\big) =$$

$$\sum_{l\geq 1}(\frac{\lambda - q(x)}{\lambda})^{l-1}\ \frac{|I(x,y)|}{\lambda} = \frac{q(x,y)}{q(x)}$$

where the two last equalities are due to

$\{W_i\}_{i\geq 1}$

are uniformly distributed in the range

$[0, \lambda]$

$q(x) < \lambda$

$|I(x,y)| = q(x,y)$

On the other hand we have

$$P(Y_n = x) = P(X_{\tau_n} = x)$$

$$= \sum_{k\geq 0} P(X_{\tau_n} = x, K_n = k)$$

$$= \sum_{k\geq 0} P(X_{S_k} = x, K_n = k)$$

So we have that:

$$P(Y_{n+1} = y \mid Y_n = x) = \frac{P(X_{\tau_{n+1}} = y, X_{\tau_n} = x)}{P(X_{\tau_n} = x)} =$$

$$= \frac{q(x,y)}{q(x)} \sum_{k\geq 0} P(X_{S_k} = x, K_n = k) \frac{1}{P(Y_n = x)}$$

$$= \frac{q(x,y)}{q(x)}$$

$$= p(x,y)$$

3.3. EXPLOSIONS

The construction made in 3.10 defines the process until the time $\tau_\infty := sup_{n \in N} \tau_n$

If I 1 = 1 then the process is defined for all time $\in [0,\infty)$. However, what if 1 <1 we have initially defined the process for times greater than 1. This section describes how to handle such cases and give a necessary and sufficient condition for x= 1.

Definition 3.3.1. Observe $\{X_t\}$defined by 3.10. We say that the process operates if

$$P(\lim_{n\to+\infty} \tau_n < \infty) > 0$$

i.e., the process explodes when an infinite number of transitions occurring in time finite. Thus, after a finite time ¿1 the process is not formally defined. Define the explosive process by adding a new call state 1 transition speeds

$$q(\infty, x) = 0$$

for all $x \in E$

The following proposition gives us information about the exponential distribution that will be useful later.

Proposition 3.3.2. Let X and Y be exponential distribution with parameters, *respectively where*

$\lambda > 0 \quad \beta > 0$

Then:

1. *Property memory loss: For s; t>0:*

$$P(X > t + s \mid X > s) = P(X > t) = e^{-\lambda t}$$

2. *Note the minimum of two numbers x and y by x ^ y. Then the random variable X ^ Y is exponential with parameter ¸+ ¯*, i.e.,

$$X \wedge Y \text{ and } \alpha + \beta$$

$$P(X \wedge Y > x) = e^{-(\lambda+\beta)}, x \geq 0$$

3. *It is true that*

$$P(X > Y) = \frac{\beta}{\lambda + \beta}$$

Suppose $\{X_n\}_{n \in N}$ *are exponentially distributed such that*

$$X_n \sim\sim Exp(\lambda_n)$$

i.e.

$$P(X_n > x) = e^{-\lambda_n x}, x \geq 0, \lambda_n > 0, n \in N$$

Then it holds that

$$\sum_{n \in N} X_n < \infty \quad \leftrightarrow \quad \sum_{n \in N} \frac{1}{\lambda_n} < \infty$$

$$P(X>t+s\mid X>s)=\frac{P(X>t+s,X>s)}{P(X>s)}$$

$$=\frac{P(X>t+s,X>s)}{P(X>s)}$$

$$=\frac{e^{-\lambda(t+s)}}{e^{-\lambda s}}$$

$$=e^{-\lambda t}$$

2. Because $\{X\wedge Y>x\}=P(X>x,Y>x)$

$$=P(X\wedge Y>x)=P(X>x,Y>x\}$$

$$=e^{-\lambda t}e^{-\beta x}$$

$$=e^{-(\lambda+\beta)x}$$

By integrating, we get

$$P(X>Y)=\iint_{\{(u,v):u>v>0\}}^{\infty}\lambda e^{-\lambda u}\beta e^{-\beta v}dudv$$

$$=\int_0^{+\infty}\beta e^{-\beta v}\left(\int_v^{+\infty}\lambda e^{-\lambda u}\,du\right)dv$$

$$=\int_0^{+\infty}\beta e^{-\beta v}e^{-\lambda v}dv$$

$$=\frac{\beta}{\lambda+\beta}$$

3. Note that if $\sum_{n\in N}X_n<\infty$ then for all s> 0 is satisfied that

$$0<E\left(e^{-sP_{n\in N}^{X_n}X_n}\right)=\prod_{n\in N}E(e^{-sX_n})=\prod_{n\in N}\frac{\lambda_n}{\lambda_n+s}$$

$$\sum_{n\in N}\left(1-\frac{\lambda_n}{\lambda_n+s}\right)=\sum_{n\in N}\frac{s}{\lambda_n+s}$$

$$\lim_{n\to\infty} \frac{s}{\lambda_n + s} = 0$$

$$\lim_{n\to\infty} \lambda_n = 0$$

$$n \to \infty$$

$$\frac{s}{\lambda_n + s} \sim \frac{s}{\lambda_n}$$

The following result allows us to state a necessary and sufficient condition for a process does not have explosions.

$$\sum_{n\in N} \frac{s}{\lambda_n + s} < \infty \leftrightarrow \sum_{n\in N} \frac{1}{\lambda_n} < \infty$$

$$\sum_{n\in N} \frac{1}{\lambda_n} < \infty$$

$$\sum_{n\in N} E(X_n) = E(\sum_{n\in N} X_n) < \infty$$

$$\sum_{n\in N} X_n < \infty$$

$$\{X_t\}_{t\in[0,\tau_\infty)}$$

Theorem 3.3.3. Given the

$\{X_t\}_{t\in[0,\tau_\infty)}$

is satisfied that *and then the process does not have explosions iff*

$$P(\tau_\infty < \infty) = P(\sum_{n\in N} \frac{1}{q(X_n)} < \infty)$$

$$P(\sum_{n\in N} \frac{1}{q(X_n)} < \infty) = 0$$

In particular it holds that if exists c> 0 such that $q(x) \leq c \; \forall \; x \in E$ then the chain has no explosions.

1. If the state space E is finite then the chain has no explosions.

Proof. Conditioned on sequence of states

$\{X_n\}$

we have to $_l$ is sum of independent random variables exponentials parameters q (X_n) that is to say

$q(X_n)$

$$\tau_\infty = \sum_{n \in N} (\tau_{n+1} - \tau_n) = \sum_{n \in N} S_n$$

where Sn are exponential distribution with parameter q (Xn). Proposition 3.3.2 then we must, conditioned on {Xn}

$$P(\ \tau_\infty < \infty) = \begin{cases} 1 \quad if \ \sum_{n \in N} \frac{1}{q(X_n)} < \infty \\ 0 \quad if \ \sum_{n \in N} \frac{1}{q(X_n)} = \infty \end{cases}$$

Taking expectations we conclude

$$P(\ \tau_\infty < \infty) = P(\sum_{n \in N} \frac{1}{q(X_n)} < \infty)$$

For (1) note that if q (x) · c for all x from E then the process does not have explosions.

On the other hand if the space E is finite then q (x) applying (1) we obtain (2).

$$\sum_{n \in N} \frac{1}{q(X_n)} \geq \sum_{n \in N} \frac{1}{c} = \infty$$

$$q(x) \leq max_{y \in E}\ q(y) = \hat{c} < \infty$$

3.4 KOLMOGOROV EQUATIONS

From now on we will use the following notations:

1. Q denote the real matrix whose entries are

$$\begin{cases} q(x,y) = \frac{p(x,y)}{q(x)} \quad if \quad x \neq y \\ q(x,x) := -q(x) = -\sum_{y \neq x} q(x,y) \end{cases} \quad in\ other \quad case$$

P_t denotes the matrix with entries

$$p_t(x, y) := P(X_t = y \mid X_0 = x)$$

3.4.1 Theorem (Chapman-Kolmogorov Equations)

$$P_{t+s} = P_s P_t \qquad (3.20)$$

t; s ¸0 is satisfied.

Proof. Using corresponding definitions and properties of statements of continuous-time Markov (homogeneous) we have:

$$c > 0 \;\; q(x) \;\leq c \; \forall x \in E$$

$$q(X_n)$$

$$\tau_\infty = \sum_{n \in N} (\tau_{n+1} - \tau_n) = \sum_{n \in N} S_n$$

$$P(\ \tau_\infty < \infty) = \begin{cases} 1 \quad if \ \sum_{n \in N} \frac{1}{q(X_n)} < \circ \\ 0 \quad if \ \sum_{n \in N} \frac{1}{q(X_n)} = \infty \end{cases}$$

$$P(\ \tau_\infty < \infty) = P(\sum \frac{1}{q(X_n)} < \infty)$$

$$\sum_{n \in N} \frac{1}{q(X_n)} \geq \sum_{n \in N} \frac{1}{c} = \infty$$

$$q(x) \leq max_{y \in E}\, q(y) = \hat{c} < \infty$$

$$\begin{cases} q(x, y) = \dfrac{p(x, y)}{q(x)} \qquad if \quad x \neq y \\ q(x, x) := -q(x) = -\sum_{y \neq x} q(x, y) \end{cases}$$

This shows that the element of Pt + s matrix is in row x and column and is obtained by the product of the row x column matrix Ps and Pt matrix, which is proved that

$$p_t(x,y) := P(X_t = y \mid X_0 = x)$$

$$P_{t+s} = P_s P_t$$

Observation 3.4.2. Note that fixed x and y in E, we can consider the function *such that*

$$p_t(x,y) := P(X_t = y \mid X_0 = x)$$

then we notice by:

$$P_{t+s} = P_s P_t$$

its derivative (with respect to t). In the following theorem give two equations LIN-lan pt0 Pt pt0 indicates where the matrix entries P0T(X, y). Before theorem see a slogan that will be useful to us.

Lemma 3.4.3. They are met:

1. lim

$$p_t(x,y) := P(X_t = y \mid X_0 = x)$$

$$P_{t+s} = P_s P_t$$

Proof. We have that

$$p_t(x,x) = P(X_t = x \mid X_0 = x)$$

$$\Sigma_{y\in E} P(X_t = y \mid X_0 = x) = 1 \quad P(X_t = x \mid X_0 = x) = 1 - \Sigma_{y\neq x} P(X_t = y \mid X_0 = x).$$

$$p_t(x,x) = 1 - \sum_{y\neq x} (tq(x,y) + o(t))$$

$$= 1 - t\sum_{y\neq x} q(x,y) - o(t)$$

$$= 1 - tq(x) - o(x)$$

$$\lim_{t\to 0}\left(\frac{1-p_t(x,x)}{t}\right) = \lim_{t\to 0}\left(q(x) - \frac{o(t)}{t}\right) = q(x)$$

and

2. $x \neq y$

$$\lim_{t\to 0}\left(\frac{p_t(x,y)}{t}\right. = \lim_{t\to 0}\frac{P(X_t = y \mid X_0)}{t}$$

$$= \lim_{t\to 0}\frac{tq(x,y) + o(t)}{t}$$

$$= \lim_{t\to 0}\left(q(x,y) + \frac{o(t)}{t}\right))$$

$$= q(x,y)$$

which completes the proof.

Theorem 3.4.4 (Kolmogorov equations). For all t>=0 are met

$P`_t = QP_t$

Q = Pt *backward equations* (3.22)

$`_t = P_t Q$

Q = Pt *Forward equatior* (3.23)

Proof.

$P`_t = QP_t$

$P`_t = P_t Q$

prove as demonstrated by reasoning analogously.

Using equations Chapman-Kolmogorov:

$$p_{t+h}(x,y) = \sum_{z\in E} p_h(x,z)p_t(z,y)$$

$$p_{t+h}(x,y) - p_t(x,y) = \sum_{z\in E} p_h(x,z)p_t(z,y)$$

$$p_{t+h}(x,y) - p_t(x,y) = \sum_{z\neq x} p_h(x,z)p_t(z,y) + p_t(x,y)$$

$$= \sum_{z\neq x} p_h(x,z)p_t(z,y) - p_t(x,y)(1 - p_h(x,x))$$

So if we see that in the conditions we are we can exchange limit with sum, we have:

$$\lim_{h\to 0} \frac{p_{t+h}(x,y) - p_t(x,y)}{h} = \lim_{h\to 0}\Big(\sum_{z\neq x} \frac{p_h(x,z)}{h} p_t(z,y) - p_t(x,y)\frac{(1 - p_h(x,x))}{h}\Big)$$

$$= \sum_{z\neq x} \lim_{h\to 0}\left(\frac{p_h(x,z)}{h} p_t(z,y) - p_t(x,y)\frac{(1 - p_t(x,x))}{h}\right)$$

$$= \sum_{z\neq x} q(x,z)\, p_t(z,y) - p_t(x,y)\frac{(1 - p_h(x,x))}{h}$$

$$= \sum_{z\neq x} q(x,z)\, p_t(z,y) - p_t(x,y)q(x)$$

$$= \sum_{z\in E} q\,(x,z)p_t(z,y)$$

where equality ultimately follows the motto 3.4.3. In this way, only to complete the test is to see that we can exchange the limit and the sum. We have fixed N

$$\lim_{h \to 0} inf \sum_{z \neq x} \frac{p_h(x,z)}{h} p_t(z,y) \geq \lim_{h \to 0} inf \sum_{z \neq x, z<N} \frac{p_h(x,z)}{h} p_t(z,y)$$

$$\sum_{z \neq x, z<N} q(x,z)\, p_t(z,y)$$

As N is arbitrary then

$$\lim_{h \to 0} inf \sum_{z \neq x} \frac{p_h(x,z)}{h} p_t(z,y) \geq \sum_{z \neq x} q(x,z)\, p_t(z,y)$$

$$N > x, \qquad p_t(z,y) \leq 1$$

Conversely if N> x, as pt (z; y) · 1:

$$\lim_{h \to 0} sup \sum_{z \neq x} \frac{p_h(x,z)}{h} p_t(z,y) \leq \lim_{h \to 0} sup \left(\sum_{z \neq x, z<N} \frac{Ph(x,z)}{h} p_t(z,y) + \sum_{z \geq N} \frac{p_h(x,z)}{h} \right)$$

$$\sum_{z \geq N} p_h(x,z) = 1 - \sum_{z<N} p_h(x,z)$$

$$= 1 - \sum_{z<N, z \neq x} p_h(x,z) - p_h(x,x)$$

$$\lim_{h \to 0} sup \sum_{z \neq x, z<N} \frac{p_h(x,z)}{h} p_t(z,y) + \frac{1 - p_h(x,x)}{h} - \sum_{z \neq x, z<N} \frac{p_h(x,z)}{h} =$$

$$\sum_{z \neq x, x<N} q(x,z) p_t(z,y) + q(x) - \sum_{z \neq x, x<N} q(x,z)$$

$$N > x$$

$$N \to \infty$$

This is true for all N> x then by N! 1 we have that

$$\lim_{N \to \infty} \sum_{z \neq x, x<N} q(x,z) = \sum_{z \neq x} q(x,z) = q(x)$$

Thus, 3.24 and 3.26 show that

$$\lim_{h \to 0} \frac{p_h(x,y)}{h} p_t(z,y) \leq \sum_{z \neq x} q(x,z) p_t(z,y)$$

3.5. CLASSIFICATION OF STATES AND INVARIANT MEASURES

In this section we some ways of classifying states of a chain of continuous-time Markov chain and define irreducible invariant measure for CMTC.

Definition 3.5.1. Be

$$T^{x \to y} := \inf\{t > \tau_1 : X_t^x = y\}$$

$$t > \tau_1$$

$$T^{x \to x} = 0$$

the first time the process begins at state x reaches the state and.

Note that in the definition asked t> 1 to prevent T x! X = 0

Definition 3.5.2. We say that the state is x from E

1. $P(T^{x \to x} = \infty) > 0$
2. $P(T^{x \to x} = \infty) = 0 \quad E(T^{x \to x}) = \infty$
3. $E(T^{x \to x}) < \infty$
4. *recurrent: if recurrent zero or positive recurrent.*

In words, a state is transient if the probability that the chain does not re-visit in the future is positive, is recurrent if the state is visited countless times with probability 1, and is recurring positive if the expected time of return to the state is finite.

Definition 3.5.3. Suppose that the process

$$\{X_t\}_{t \geq 0}$$

is irreducible if for all x from E and the probability of reaching the state and out of state x in a finite time is positive, i.e., if

$$P(T^{x \to y} < \infty) > 0$$

Below we give definitions chains invariant measure for discrete time and continuous time.

Definition 3.5.4. Observe $\{X_n\}_{n \in N}$ a Markov chain with discrete time space E numerable states and transition matrix Q that is

$x, y) = P(X_1 = y \mid X_0 =$

and

$(Q: E \times E \to [0,1] \quad \sum_{y \in E} Q(x, y) = 1)$

A probability measure [1] defined in E is called invariant with respect to Q if verified that:

$$\mu(x) = \sum_{z \in E} \mu(z) Q(z, x)$$

$$\pi Q = 0 \qquad (3.29)$$

Definition 3.5.5. We say that a probability measure in E ¼ call is invariant or stationary continuous process for fXtgt¸0 if verified that:

$$\sum_{x \in E} \pi(x) q(x, y) = 0 \quad \forall y \in E$$

$$\sum_{x \in E} \pi(x) = 1$$

where Q is defined at the beginning of section 3.4 matrix.

3.6. SKELETON PROCESS

Let's assume in this section that the Xn process and its skeleton

$$\{Y_n\}_{n \in N}$$

defined by 3.2.3 are irreducible and positive recurrent. No. We know that a measure is invariant for discrete $\{Y_n\}_{n \in N}$ chain if it satisfies

$$\sum_{x \in E} \pi(x) = 1$$

$$\sum_{x \in E} v(x) p(x, y) = v(y)$$

On the other hand, a ¼ measure is invariant for continuous process if it satisfies

$$\sum_{x \in E} \pi(x)q(x,y) = \pi(x)q(y)$$

The following result shows the relationship between an invariant measure for discrete and process invariant measure for the continuous process.

Proposition 3.6.1. For a probability measure on E, and consider the measure

$$\pi(x) := \frac{v(x)}{q(x)}\left(\sum_{z \in E} \frac{v(z)}{q(z)}\right)^{-1}$$

$$\{X_t\}_{t \geq 0}$$

$$\{Y_n\}_{n \in N}$$

Then it holds that it is invariant to $\{X_t\}_{t \geq 0}$*if and only if the measure is invariant ¼ process for* $\{Y_n\}_{n \in N}$.

Proof. Suppose $\{Y_n\}_{n \in N}$ is invariant, then

$$v(y) = \sum_{x \in E} v(x)p(x,y)$$

$$= \sum_{x \in E} v(x)\frac{q(x,y)}{q(x)}$$

Then

$$\sum_{x \in E} \frac{v(x)}{q(x)} q(x,y)\left(\sum_{z \in E} \frac{v(z)}{q(z)}\right)^{-1} = v(y)\left(\sum_{z \in E} \frac{v(z)}{q(z)}\right)^{-1}$$

$$\sum_{x \in E} \pi(x)q(x,y) = \pi(y)q(y)$$

which, applying the definition of the measure it is equivalent to ¼

$$\sum_{x \in E} \frac{v(x)}{q(x)}\left(\sum_{z \in E} \frac{v(z)}{q(z)}\right)^{-1} q(x,y) = \frac{v(y)}{q(y)}\left(\sum_{z \in E} \frac{v(z)}{q(z)}\right)^{-1} q(y)$$

$$\sum_{x \in E} v(x)p(x,y) = v(y)$$

and therefore it is invariant. Conversely, we know that

$$(x) = q(y)$$

$$x, y \in E$$

$$\pi(x) = v(x)\ \forall x \in E$$

$$\{Y_n\}_{n \in N}$$

then using the definition of the measure have to ¼

$$\sum_{x \in E} \frac{v(x)}{q(x)} \left(\sum_{z \in E} \frac{v(z)}{q(z)} \right)^{-1} q(x, y) = \frac{v(y)}{q(y)} \left(\sum_{z \in E} \frac{v(z)}{q(z)} \right)^{-1} q(y)$$

Simplifying we get that

$$\sum_{x \in E} v(x) p(x, y) = v(y)$$

and therefore it is invariant for the process.

Corollary 3.6.2. If q (x) = q (y) for all x; and then

$$x, y \in E$$

$$\pi(x) = v(x)\ \forall x \in E$$

$$\{Y_n\}_{n \in N}$$

Proof. Let q (x) = q for all

$$x \in E$$

we have

$$q(x) = q \in R$$

$$\pi(x) = \frac{v(x)}{q} \left(\sum_{z \in E} \frac{v(z)}{q} \right)^{-1}$$

$$= v(x) \left(\sum_{z \in E} v(z) \right)^{-1} = v(x)$$

$$v(x) = \pi(x)\ \forall\ x \in E$$

3.7. BIRTH AND DEATH PROCESSES

Processes of birth and death represent growth or extinction of the population of a certain system. The value of the variable X_t. It represents the number of individuals in the system at time t. The birth and death rates depend only on the number of individuals there at the time, i.e.:

$$q(x, x+1) = \lambda_x^+ \quad q(x, x-1) = \lambda_x^-$$

where these are families of non-negative parameters. Next we will use balance equations to find conditions under which the process permits an invariant measure.

As the state space E is countable we assume that N is for notational convenience. We find a vector $\pi(x)$, g in N that satisfies the following conditions:

$$\pi(0) \sum_{z \in N} q(0, z) = \sum_{y \in N} \pi(y) q(y, 0)$$

$$\pi(x) \sum_{z \in N} q(x, z) = \sum_{y \in N} \pi(y) q(y, x)$$

Substituting in each case the corresponding transition speeds 3.34 and 3.35 imply that

$$\pi(0) q(0,1) = \pi(1) q(1,0)$$

$$\pi(x)[q(x, x+1) + q(x, x-1)] = \pi(x-1) q(x-1, x) + \pi(x+1) q(x+1, x)$$

$$\pi(x+1) = \pi(x) \frac{\lambda_x^+}{\lambda_{X+1}^-} \forall x \geq 0$$

and

$$\pi(x) = \prod_{i=0}^{x-1} \frac{\lambda_i^+}{\lambda_{i+1}^-} \pi(0) \quad \forall x \geq 1$$

$$\pi = \{\pi(x)\}_{x \in N}$$

$$\sum_{x \in N} \pi(x) = 1$$

$$\sum_{x \in N} \pi(x) = 1$$

$$\pi(0) > 0$$

Thus, by building it holds that ¼ = f¼ (x) g_{x2N} verifies the equations balance. Lack asking him to meet ¼(X) = 1. This requires that ¼ (0)> 0, then

$$\sum_{x\geq 1}\frac{\lambda_0^+\lambda_1^+\dots\lambda_{x_1}^+}{\lambda_0^-\lambda_1^-\dots\lambda_{x_1}^-}<\infty$$

$$\pi(0):=(\sum_{x\geq 1}\frac{\lambda_0^+\lambda_1^+\dots\lambda_{x_1}^+}{\lambda_0^-\lambda_1^-\dots\lambda_{x_1}^-})^{-1}$$

and $\pi(\mathrm{x})$ inductively.

CHAPTER 4

APPLICATIONS IN BIOLOGY

CONTENTS

All biological systems share ownership of the growth of its indivisible UOS therefore when studying populations (for example plants or animals) the existence of a theory about growth processes is extremely necessary. In this chapter we will see some applications the growth of populations Stochastic processes studied in previous chapters of this work possess.

4.1. MODELING THE GROWING POPULATIONS

Processes of birth and death that are more general were studied by Kendall [7], in which birth and death rates are arbitrary functions of time and nonlinear functions of the variable states as we have assumed so far.

4.1.1. Birth and Death Processes

In this section we will begin considering birth and death processes, which are processes FXT: t, 0g where the random variable Xt indicating certain population size at time t may experience positive and negative jumps. The characteristics of these types of processes are:

1. Ifintime*t*thesystemisinstatex(x=1,2,::)transitionprobabilitytostate x + 1 in the time interval $\{X_t\colon t \geq 0\}(x = 1,2,\ldots)$ $(t, t + \boldsymbol{\Delta t})$ $\lambda_x \boldsymbol{\Delta t} + \boldsymbol{o}(\boldsymbol{\Delta t})$ is say

$P(X_{t+\Delta t} = x + 1 | X_t = x) = \lambda_x \boldsymbol{\Delta t} + \boldsymbol{o}(\boldsymbol{\Delta t})$

2. If in time *t* the system is in state x (x = 1, 2,: :) transition probability to state gig 1 in the time interval

$(t, t + \boldsymbol{\Delta t})$

$\mu_x \boldsymbol{\Delta t} + \boldsymbol{o}(\boldsymbol{\Delta t})$

is say:

3. $P(X_{t+\Delta t} = x - 1 | X_t = x) = \mu_x \boldsymbol{\Delta t} + \boldsymbol{o}(\boldsymbol{\Delta t})$

4. The probability of transition from one state to another than neighbor is or (Dt) i.e., :

$P(X_{t+\Delta t} = y\, | X_t = x) = \boldsymbol{o}(\boldsymbol{\Delta t}) \quad \boldsymbol{y} \notin \{\boldsymbol{x} - \boldsymbol{1}, \boldsymbol{x} + \boldsymbol{1}\}$

The probability that no change in the process state is 1¡ (x + ¹x) +.Δ.t or (.Δ.t).

4. X = 0 state is absorbent

where the coefficients $\lambda_x \mu_x$ are state functions x. Note

$$P_x(t) := P(X_t = x)$$

then the above properties imply that:

$$P_x(t+\Delta t) = \lambda_{x-1} P_{x-1}(t)\Delta t + \mu_{x+1} P_{x+1}(t)\Delta t + (1-(\lambda_x+\mu_x)\Delta t)P_x(t) + o(\Delta t)$$

From this relation it follows easily that for all x = 1; 2;...

$$\frac{\partial P_x(t)}{\partial t} = \lambda_{x-1}(t)P_{x-1}(t) + \mu_{x+1}P_{x+1}(t) - (\lambda_x+\mu_x)P_x(t)$$

$$\frac{\partial P_0(t)}{\partial t} = \mu_1 P_1(t)$$

$$\lambda_{-1} = \lambda_0 = \mu_0 = 0$$

since 1 = 0 = 0 = [1]0 since the state 0 is absorbent. If we assume that at time zero the system is in state x = x0 with 0 <x0 <1 then the initial conditions of the system are given by:

$$x = x_0$$

$$0 < x_0 < \infty$$

$$P_x(0) = \delta_{x,x_0} = \begin{cases} 1 & if \quad x = x_0 \\ 0 & if \quad x \neq x_0 \end{cases}$$

$$P_x(t) \tag{4.1}$$

Below we shall find the distribution $P_x(T)$ for the particular case of birth and death processes simple or linear in which $\lambda_x = \lambda x$ for all x in N where x> 0. To this end, we introduce the generating function

$$\lambda_x = \lambda x$$

$$\mu_x = \mu x$$

$$\mu, \lambda > 0$$

$$F(s,t) = \sum_{x=0}^{\infty} P_x(t)s^x \tag{4.2}$$

Equations 4.1 and 4.2 is easy to see that the above generating function satisfies the following differential equation:

$$F(s,t) = f\left(\frac{\mu-\lambda s}{1-s} e^{-(\lambda-u)t}\right) \quad (4.3)$$

$$X_0 = x_0 = 1$$

The general solution is

$$F(s,0) = s \quad (4.4)$$

where f (¢) is an arbitrary function. If we assume that $X_0 = x_0 = 1$ then we have F(s, 0) = so is

$$s = f(\frac{\mu - \lambda s}{1 - s})$$

and thus

$$f(\varsigma) = \frac{\mu - \varsigma}{\lambda - \varsigma} \quad (4.5)$$

Then we obtain

$$F(s,t) = \frac{\mu(1 - e^{(\lambda-\mu)t}) - (\lambda - \mu e^{(\lambda-\mu)t})s}{\mu - \lambda e^{(\lambda-\mu)t} - \lambda(1 - e^{(\lambda-\mu)t})s} \quad (4.6)$$

If expanded as a series in powers of s, sx coefficients give Px (t) which is what we want to find. So:

$$P_x(t) = (1 - \alpha(t))(1 - \beta(t))[\beta(t)]^{x-1} \quad (4.7)$$

$$P_0(t) = \alpha(t)$$

$$\alpha(t) = \frac{\lambda - (e^{(\lambda-\mu)t} - 1)}{\lambda e^{(\lambda-\mu)t} - \mu} \quad (4.8)$$

Usually in applications processes of birth and death are not the probabilities Px (t) but the most interesting moments of Xt and the probability of extinction of the population. Then we get differentiating the generating function and B.1.1 and B.1.2 proceeding according to the following expressions for the expectation and variance of Xt variable:

$$E(X_t) = e^{(\lambda-\mu)t}$$

$$var(X_t) = \frac{\lambda + \mu}{\lambda - \mu} e^{(\lambda-\mu)t}(e^{(\lambda-\mu)t} - 1) \quad (4.9)$$

Thus we see that:

$$\lim_{t\to\infty} E(X_t) = \begin{cases} 0 & if \quad \lambda < \mu \\ 1 & if \quad \lambda = \mu \\ \infty & if \quad \lambda > \mu \end{cases} \tag{4.10}$$

which tells us that if the rates of death and birth ,; [1] are equal, then the expected rate of population growth is zero and the size of the population is stationary.

Another important fact is clear from the above results is the probability P0 population extinction (t), i.e., the probability of the population "develop-look" at time t:

$$P_0(t) = \alpha(t) = \frac{\mu(e^{(\lambda-\mu)t}-1)}{\lambda e^{(\lambda-\mu)t}-\mu}$$

Then the probability that the population becomes extinct at some point is:

$$\lim_{t\to\infty} P_0(t) = \begin{cases} 1 & if \quad \lambda < \mu \\ \frac{\mu}{\lambda} & if \quad \lambda > \mu \end{cases}$$

and then we see that naturally thinking, the population shall be extinguished with probability 1 if the rate of deaths higher than the birth rate.

4.1.2. Birth and Death Processes That Are Not Homogeneous

Processes of birth and death more general were studied by Kendall [7], in which birth and death rates are arbitrary functions of time and nonlinear functions of the variable states as we have assumed so far. These processes are characterized by the following differential equation:

$$\frac{\partial P_x(t)}{\partial t} = \lambda(t)(x-1)P_{x-1}(t) + \mu(t)(x+1)P_{x+1}(t) - \big(\lambda(t)+\mu(t)\big)xP_x(t)$$

for all x=1,2,...

$$\frac{\partial P_0(t)}{\partial t} = \mu(t)P_1(t)$$

where

$$P_x(t) = P(X_t = x)$$

with x 2N and Xt denotes the size of the population at time t. Analogously to the reasoning made in the previous section is that the solution of this system with initial condition for x = 1; 2;..., Y

$$\frac{\partial P_0(t)}{\partial t} = \mu(t)P_1(t)$$

$$P_x(t) = P(X_t = x)$$

where

$$P_x(0) = \begin{cases} 1 & if \quad x = 1 \\ 0 & if \quad x \neq 1 \end{cases}$$

$$P_x(t) = (1-\alpha)(1-\beta(t))[\beta(t)]^{x-1}$$

$$P_0(t) = \alpha(t)$$

$$\alpha(t) = 1 - \frac{e^{-\gamma(t)}}{w(t)}$$

$$\beta(t) = 1 - \frac{1}{w(t)}$$

$$\gamma(t) = \int_0^t (\mu(\tau) - \lambda(\tau))d\tau$$

$$w(t) = e^{-\gamma(t)}\left(1 + \int_0^t \mu(\tau)\, e^{\gamma(\tau)} d\tau\right)$$

As stated earlier, the interesting thing about these processes is the behavior of the moments and the probability of extinction of the population. In the same manner as was done in the previous section, it is not difficult to see that:

$$E(X_t) = e^{-\gamma(t)}$$

$$var(X_t) = e^{-2\gamma(t)} \int_0^t (\lambda(\tau) + \mu(\tau))e^{\gamma(t)} dr$$

where it is given , (t) [1] manner (t) we know explicitly the expressions for the expectation and variance of variable X_t and study their comparison, such as linear case study in the previous section. Furthermore, due to the definition of ® (t) we see that in this general case the probability of population

$P_0(t) = \frac{\int_0^t \mu(\tau)e^{\gamma(t)}d\tau}{1+\int_0^t \mu(\tau)e^{\gamma(\tau)}d\tau}$ it is terminated at time is:

$$P_{x,y}(t) := P(X_t = x, Y_t = y\} \quad \forall x = 0,1, \ldots$$

Note that in the case of x0 = X0> 1 independence between individuals of the population implies that the probability of population extinction is obtained by raising 4.10 to x0-th power. Now consider the "cumulative process" associated with a process-to Birth and death. Thus, besides the Xt is representing the number of individuals in the population at time t, we introduce a new variable representing Yt total number of births that occurred in the population until time t. Yt is a process of "pure birth" in the sense that a change in Yt is induced by a source in the population, i.e., by an increase in one unit in the random variable Xt.

Let $P_{x,y}(t)$ be a cumulative process, and note

$$P_{x,y}(t) := P(X_t = x, Y_t = y\} \quad \forall x = 0,1,...$$

In order to study the bivariate above process we introduce the generating function

$$G(r,s,t) = \sum_{x=0}^{\infty}\sum_{y=0}^{\infty} P_{x,y}(t) r^x s^y \quad |r| \leq 1, |s| \leq 1$$

$$\{X_t, Y_t : t \geq 0\}$$

and either F (r, t) generating function associated fXtgt,0 process. We know from the previous section that verifies

$$\frac{\partial F(r,t)}{\partial t} = (\lambda r^2 - (\lambda + \mu)r + \mu)\frac{\partial F(r,t)}{\partial r}$$

and noting that Y_t only shares with X_t "positive" jumps, it is proved that:

$$\frac{\partial G(r,s,t)}{\partial t} = (\lambda r^2 s - (\lambda + \mu)r + \mu)\frac{\partial G(r,s,t)}{\partial r} \qquad 4.11$$

given that the objective of this chapter is to present some applications of the theory described in the previous chapters, we Then the solution of the differential equation and some other results for the particular case where the speeds of birth and death are constant and when the initial conditions are

$$X_0 = Y_0 = 1.$$

We are interested in finding the moments then Xt and Yt of the variables, and for that introduce the "cumulative" generating function

$$K(u,v,t) = \log G(e^u, e^v, t)$$

From Equation 4.11 it follows that:

$$K(u,v,t) = uE(X_t) + vE(Y_t) + \frac{u^2}{2}var(X_t) + uvCov\{X_t, Y_t\} + \frac{v^2}{2}var(Y_t) + \cdots$$

and in other hand we have

$$\frac{\partial K(u,v,t)}{\partial t} = (\lambda(e^{u+v}-1) - \mu(1-e^{-u}))\frac{\partial K(u,v,t)}{\partial u}$$

Expanding both sides of the equation in powers of the previous equation coefficient equaling expressions that relate the moments of the variables are obtained, and solving these equations we obtain:

$$E(X_t) = e^{-\gamma t}$$

$$var(X_t) = e^{-2\gamma(t)} \int_0^t (\lambda(\tau) + \mu(\lambda))e^{\gamma(\tau)}\, d\tau$$

and

$$E(Y_t) = 1 + \int_0^t e^{-\gamma(\tau)}\lambda(\tau)d\tau$$

$$Cov\{X_t, Y_t\} = e^{-\gamma(\tau)} \int_0^t (1 + \frac{var(X_\tau)}{E(X_\tau)})\lambda(\tau)d\tau$$

$$var(Y_t) = \int_0^t [E(X_t) + 2Cov\{X_t, Y_t\}]\, \lambda(\tau)d\tau$$

For the special case where

$$\lambda(t) = \lambda$$

and

$$\mu(t) = \mu$$

know from the results obtained in the first section of this chapter is the limit behavior as the first moments of the variable X_t. From the above equations we can also study these moments for the new variable Y_t with the following results:

$$\lim_{t\to\infty} E(Y_t) = \frac{\mu}{\lambda - \mu}$$

$$\lim_{t\to\infty} var(Y_t) = \frac{\lambda\mu(\lambda+\mu)}{(\mu-\lambda)^3}$$

$$\lim_{t\to\infty} Cov\{X_t, Y_t\} = 0$$

4.1.3. Simulating Birth, Death and Immigration Processes

The type of birth and death process that is of interest in various applications is the process of birth, death and immigration. The characteristics of these processes are the same as those of the simple process of birth and death, with the additional property that the time interval (t, t + t) the probability that the size of the population increases one unit due to immigration from outside the population is th t + o (at), where th > 0 is the rate of immigration. Note that although immigration increases the size of the population, the effect of the rate of immigration differs from the rate of births , in the sense that it is independent of the size of the population.

Let $F_I(S, t)$ the generating function associated with this process. Then it is true:

$$\frac{\partial F_I(s,t)}{\partial t} = (\lambda s^2 - (\lambda + \mu)s + \mu)\frac{\partial F_I(s,t)}{\partial s} + v(s-1)F_I(s,t) \qquad (4.12)$$

where F (s, t) is the generating function associated with the birth and death process. If X0 = 0 then 4.11 or 4.12 will be solved with the initial condition FI (s, 0) = 1. The solution of 4.12 is:

$$\frac{\partial F_I(s,t)}{\partial s} = vF_I(s,t)(F(s,t) - 1)$$

$$F_I(s,t) = (\frac{\lambda - \mu}{\lambda e^{(\lambda-\mu)t} - \mu})^{v/\lambda}[1 - s\left(\frac{\lambda e^{(\lambda-\mu)t} - 1}{\lambda e^{(\lambda-\mu)t} - \mu}\right)]^{-v/\lambda} \qquad if \quad \lambda \neq \mu$$

Differentiating the above equations we obtain the following expression for the expectation of population size at time t:

$$F_I(s,t) = (1 + \lambda t - \lambda ts)^{\frac{v}{\lambda}} \qquad if \quad \lambda = \mu$$

$$\frac{\partial F_I(s,t)}{\partial t}]s = 1 = E(X_t) = \begin{cases} \dfrac{v}{\mu - \lambda} & if \quad \lambda \leq \mu \\ \infty & if \quad \lambda \geq \mu \end{cases}$$

4.1.4. Modeling the Growth of Population by Gender

In the previous sections we have restricted our attention to growth consist of a single class of individuals. In many problems that are of interest in the study of human population growth, for example, is not the total number of individuals that is of interest principal but the growth of each sex in the general population. This section presents stochastic models for Plant growth-population according to sex.

Are random variables and respectively represent the number that female and male population is in a certain time t, and is

$$P_{x,y}(t) := P(X_t = x, Y_t = y)$$

with x; and , 0 the joint distribution of and . We assume that the population is developed according to the following mechanism:

Populations are generated by two independent individuals.

A living female time t has probability pt+Δt to reproduce another female and probability qt+Δt playing a male in the range of time (t, t+Δt), where t and p+q=1.

A living female time t is likely died in the range (T, t+Δt) and a living male time t i.e., in said time interval.

Of the above conditions it is that the stochastic process of growing population by sex is a bivariate type process birth and death. Considering the case where time t = 0 the population is comprised of only one female and lock the same manner as in previous sections (using generating functions) differential equations that link the different times are obtained, and solving these equations we can easily explore what the asymptotic behavior of these moments [9]. Now the interesting thing in these cases is also studying the asymptotic behavior com-the ratio of the expected number of males and females of the population. Given then that expressions obtained for the first moments of the variables are:

$$E(X_t) = e^{(\lambda p - \mu)t}$$

$$E(Y_t) = \frac{\lambda q}{\lambda p - \mu + \mu`}(e^{(\lambda p - \mu)t} - e^{-\mu` t})$$

it is then that in the case where

$$(\lambda p - \mu + \mu`) > 0$$

$$\lim_{t\to\infty} \frac{E(X_t)}{E(Y_t)} = \frac{(\lambda p - \mu + \mu`)}{\lambda q}$$

Thus if the probability of death in females equal to that of males, then:

$$\lim_{t\to\infty} \frac{E(X_t)}{E(Y_t)} = \frac{p}{q}$$

Another point that may be interesting in this model is to consider the distribution of females in the population. The characteristics of the definition of this process and the expression of the expected number of females in the population is that the growth of females follow a simple process of birth and death. Be

$$P_x(t) := P(X_t = x) = \sum_{y=0}^{\infty} P_{x,y}(t) \qquad x = 0,1,\dots$$

Differential equations valid for Px,y (t) (See ref Bharucha-Reid) is obtained that:

$$\frac{\partial P_x(t)}{\partial t} = -(\lambda p + \mu)xP_x(t) + \lambda p(x-1)P_{x-1}(t) + \mu(x+1)P_{x+1}(t) \quad x = 1,2,\dots$$

The initial conditions are

$$\frac{\partial P_0(t)}{\partial t} = \mu P_1(t)$$

$$P_x(0) = \begin{cases} 1 & if \quad x = 1 \\ 0 & if \quad x \neq 1 \end{cases}$$

$$\lim_{t\to\infty} P_0(t) = \begin{cases} 1 & if \quad \lambda < \mu \\ \frac{\mu}{\lambda} & if \quad \lambda > \mu \end{cases}$$

The theory presented in the first two sections of this chapter we know that:

$$P_x(t) = (1-\alpha(t))(1-\beta(t))[\beta(t)]^{x-1} \qquad x = 1,2,\dots\text{where } P_0(t) = \alpha(t)$$

It is then the probability of population extinction is:

$$\alpha(t) = \frac{\mu(e^{(\overline{\lambda p}-\mu)t}-1)}{\lambda p\, e^{(\lambda p-\mu)t}-\mu}$$

$$\beta(t) = \frac{\lambda p(e^{(\lambda p-\mu)t}-1)}{\lambda p e^{(\lambda p-\mu)t}-\mu}$$

The models that we have been describing along the sections of this statement-what are some important models when studying population growth, but there are many more. Besides these, growth models population of great importance in ecology for example are those described growth constituted by two or more classes of individuals who interact with each other (see for example [10] or [11]) or population migration models in which individuals can migrate from one region to another. There are also processes that shape the development of certain epidemic in a population, or models for the growth of heterogeneous populations where one type of organism can "mutate" into a different organism. Generally there are many applications that the continuous time stochastic processes are in bi-ology but these applications are not restricted only to this field, but these processes are also helpful in physics, chemistry, queuing theory, etc. For most applications we cite for example [9].

CHAPTER 5

NETWORK ANALYSIS MODELED WITH STOCHASTIC PROCESSES

CONTENTS

5.1. INTRODUCTION

Service centers with networked care systems live day-to-day problems of congestion, inefficiency and lack of planning in the proper design of their model of care. Both conception and redesign involves a study based on performance measures that are affected by many variables constantly in and out of the system.

Each of these variables impact somehow the way the system behaves and how it is able to meet customer needs efficiently. Especially the human resource is the key factor in good design because it has a greater impact on measures of assessment and therefore represent the best starting point for implementing process improvement, study and analysis is only possible through statistical tests demonstrating their effect on the indicators around interactions with other factors affecting a queuing model.

This study is responsible for conducting this analysis combining different types of factors such as attendance rates arrival, probability distributions, length of a serial network, number of servers and system capacity to model queuing problems in different applications and significantly different results.

In addition from the obtained analysis, it is possible to construct an optimization algorithm that involves knowledge of the factors and their influence on indicators, so it's wise in choosing the best set of solutions to the problem, i.e., the number of servers and the capacity to meet customer needs. Techniques multi-objective optimization is very useful when minimize competing goals where there are no parameters that optimize both objectives simultaneously.

The inclusion of techniques known in operations research an optimization algorithm conceived from game theory, developed processes local search and evaluated by economic principles of equality, provides an analysis tool untraditional and innovative with different conclusions about the optimal state of a system.

Thus a research study and analysis of queuing networks that provide criteria improvement and efficiency in the efficient allocation of human resources is conceived.

5.2. STATEMENT OF THE PROBLEM

5.2.1. Background

The significant increase in demand for services created within organizations growth and improvement as part of their corporate dynamic evolution. Likewise, the increase in profits offers the possibility to invest in themselves with the aim to extend and expand its influence in the market to which they belong. Despite this, the problems of inadequate planning resources for meeting the needs of its customers continue and indeed intensify as the organization grows. Congestions and waiting situations are presented to the extent that the service offering is inferior to the demand for it and called tails, which also annoying, subtract points to the perceived quality of service are created.

The study of waiting lines or queues began in 1909 when AK Erlang, Danish engineer, analyzed telephone traffic systems where waiting calls to be switched to its destination, but given the growth of the service, waiting to be communicated. It increased and sometimes it was impossible to complete the operation. From this, an investigation Erlang formalized after be known as queuing theory in order to analyze the expected and the factors involved in it.

Although the idea of queuing theory emerged from a network, only it was conceived, worked and determined as such in studies by Jackson fifty years later, where the open and closed queuing networks were analyzed and showed Check rates following the Markovian property are independent of network status. Under a mathematical proof the study concluded that given this networks are open when there arrivals entities outside the system from the network. In closed networks nodes form a cycle that never allows the flow of entities outside of it.

It is equivalent to setting a conditional probability of any "event" future given any "event" past and present state $X_i = I$, is independent of the past event and depends only on the current state of the process. This property was possible to analyze the network as a set of individual and independent systems expected rate.

This finding positively impact the theoretical development of networks expected to analyze complex systems allowing reflecting the true structure of the systems and fit in better as a reality. In 1962, Mazumdar, Mason and Douligeris networks Jackson used to represent models telecommunications and determine the maximum capacity of these systems before collapse. In turn, PJ Burke 1976 studied the behavior of queues M/M/1 into a network of Jackson and analyzed how the outputs of a process to another behaved equivalently with the distribution of arrival whose arrivals conformed to distributions classical probability.

The importance of studying Jackson has had a high impact on queuing theory and its contribution has transpired over the years. Down authors in their investigation as 1998, worked with the combination of open and closed Jackson networks to determine the stability of recurrence and transition from one system to the extent entering, leaving or entities remained inside it. Other examples are evident in the work carried out in 2008 by Gonga and Shouyang who use closed networks Jackson to model a supply chain in which the number of servers and workstations should be determined to find the balance of the system and does not generate necks bottle.

Thus developed to that point in the classical theory of waiting lines, this analysis became descriptive rather than prescriptive; In practice, this meant that a given probability distributions of inter-arrival times and service times, average performance is quantified, but these did not translate into strategies or decisions regarding optimal design system. By this very fact seemed logical to use queuing theory not only as a method of analysis, but as a tool for decision, which was made possible by integrating theory with optimization. JS works such as bars and DJ Makowski conducted in 1985, queuing theory used to plan long term and which would be useful productivity of a system of queues k; to find the optimum number of servers to change them over time. Meanwhile, Ann Van Mieghem in its service-oriented application, focused on the use of queuing theory as an input for the same cost optimization to determine whether or not to increase the capacity of the system, because they knew that regardless of the service, queues were more costly in terms of quality if there were people who needed attention on them.

Some applications of the business field, besides needing to be optimized, have a complex structure and configuration that does not allow them traditional solution using mathematical methods described herein to. The vast majority of networks, for example, are complicated systems are difficult to handle, whose probability distributions used to model certain phenomena do not necessarily follow an exponential behavior; or where certain restrictions, such as capacity, make it impossible to mathematical resolution. This is where comes and simulation as a tool integrated to provide solutions to these issues and complement the theoretical development with practical.

In addition, refinement and use of computers gave a decisive boost to the development and use of simulation. Jamhoura, Pennaa, Nabhenb and Pujolle found in simulating the response of how to reduce queues data on computer networks. In this study predicting many scenarios the system was to later compare the results and find the most efficient. This is one of many examples where the hybrid optimization/simulation was instrumental in treating problems queue until then because of the difficulty in mathematical modeling had not been treated. Investigating cases like Jeung Hyun, Chang and Taioun on logistics service providers (3PLs) 13. It shows how through an optimization model these contractors determine the structure of its service network in the provision of services (improved times, distribution and storage) to companies with high quality standards. Later in this same study 3PLs uses a simulation model in order to predict the behavior and performance measurement system one of its customers.

Vanda, Felici and Impelluso explored new applications in the design and management of a health center. In this study a set of services are required by one or more categories of users and defining the number of servers assigned to each service affects overall system efficiency and indicators. The main contribution of the research is the interactive methodology that uses simulation to estimate the objective function that calculates and validate the optimal server configuration. The development of this tool is to support health managers in making decisions not leave to chance the increase in the number of servers and to determine the effects this has on system performance and budget plan.

Meanwhile, M. Kuroda and A. Kawada considered the ability of a generalized system sales using optimization minimized the average waiting time, and the simulation proved the effectiveness of measures taken under horizons where demand or the arrival rates varied system unexpectedly.

On the latter subject, they have also been studied where the behavior of arrivals in the provision of services does not follow a uniform pattern as analyzed so far, or instead have changes in arrivals, low distributions do not necessarily exponential. In 2007, Jeongsim Kim and Kim Bara conducted research on systems with a single server where arrivals were governed by regular variations in their rates. In these systems, it was shown how the size of the queues grew rapidly in some time periods while in others it generated no waiting. In turn, Feldman, Mandelbaum, Massey and Whitt addressed in 2008 varying rates arrival at a call center and developed a method, Flexible via simulation based on an iterative algorithm-staffing, to determine appropriate levels of care. The greatest contribution of this research is the finding of a stable point of care, despite uncertainty over rates arrival.

Modeling systems via queuing networks and its solution through optimization tools and/or simulation is used in different application fields such as control hydroelectric systems design chains abastecimiento, traffic optimization, sales estimate, traffic control, performance evaluation service centers, among others; where they have complex structures that require strategic decisions that improve system performance and positively impact the perceived quality of service. These industrial applications are some of the examples used by Dembo, Mulvey and Zenios as prototypes or models generalized by its versatility and flexibility could be used to optimize time, costs, and generally increase the productivity of different applications in networks. However, even they not have been involved in the study of these applications process variations arrivals, which far from always random, cyclical fluctuations may occur, as does the demand for example, along the time.

In addition to the aforementioned aspects, for the optimization of enterprise applications also we need to consider the efficient allocation of resources that add value within these processes, because this factor impacts the performance and productivity of the systems. Related to this aspect, Hillier and So analyzed the behavior of a serial network in order to efficiently allocate resources in the process. The purpose of the optimal allocation of servers within the network was to minimize production times and costs of processing parts in the manufacturing line. The methodology used for this purpose started from an initial allocation, and on this modified configuration of the workforce in each node to find that combination where a machine increased significantly improved performance measures established for the system. This procedure was called L-phenomenon.

Other authors interested in minimizing waiting times within a system, applied nonlinear optimization models for the efficient allocation of servers. For Berman, Larson and Chiu, based on some properties of location servers, sought the optimum use of resources depending on the type of demand: the type of service demand determined a different allocation of available servers (location and quantity).

Meanwhile, Marianov and more recent studies analyzed how an alternative servers distribute different kinds (for example office, ATM and machines, among others) in service networks. The main objective was to reduce the impact of queues on users, i.e., minimize waiting times for enterprise applications such as banks, hospitals and box office sales. For the definition of the optimal distribution using a new heuristic called maximal covering, with which, in addition to solving the initial problem posed, the importance of the amount allocated in the first nodes of the network servers evidenced raised.

After analyzing the behavior and performance of queuing networks in service processes, how to minimize the costs associated with waiting and residence times in the system to efficiently allocate human resources in organizations such structures?

5.3. JUSTIFICATION OF USING STOCHASTIC PROCESSES

The service sector every day is becoming one of the axes of major development of the national economy, which is evident in its contribution to the Gross Domestic Product (GDP) and job creation in the labor market, more than any other sector of the economy. DANE statistics show that in 1980 the sector accounted for 47% of GDP, a share that rose to 59% in 200731 and remained in the third quarter of 2008 with growth of 20.1% over the period previous. Sector growth positively impacts the contracts signed, which is evident in the increase of 7.8% in the first quarter of 2008 compared to January–March 2007.

In light of these figures, it is possible to infer that the sector is one of its best moments. It is vital to analyze their behavior with the aim of generating innovative ideas to maintain/potentiate growth, create wealth for the community and consolidate economy. Banks, hospitals, customer service and other organizations should focus their attention on points such as development and product design, pricing structure, promotion strategies and the definition of places and distribution channels. But in addition to this, service companies should focus more closely on staff in the process as such and the physical evidence of the service, as set by the marketing services in the marketing plan.

However there is a common problem that prevents the efficient management of an organization through its human talent, processes and physical evidence of the service and are precisely the queues. According to Michael Fortino, Priority Management in Pittsburgh, most people spend about five years of his life in ranks, a phenomenon that becomes more prevalent as long as demand for services increases more than proportionally with respect to their ability to offer, causing boredom, discomfort, dissatisfaction and costs. No organization is immune to this problem, in fact any request, payment advice, claim or acquisition of products are only examples of how daily cola consumers should expect to form in front of a service. At airports like Atlanta and Chicago, for example, in 1995 the costs for airlines by the tails were quantified around $6 to $10 per minute, which accumulated in a year represent 10 million dollars, plus many other intangible costs and hidden as the deterioration of corporate image and quality of service are difficult to gauge and quantify. The time the population of a country loses to wait in queues is an important factor in the quality of life and the efficiency of its economy, because by investing downtime waiting in productive activities could generate economic benefits country and reducing organizational costs. In world powers like the United States, globally companies are paying more in staff costs due to increases in the value of the hand and overstaffing in Europe, the United States, Japan and especially in developing economies such as Eastern European. Accordingly, know for sure how many employees should be assigned to a work order to meet service standards becomes a major factor. This planning should be given primarily to the customer, because it is the main profit generator and is to whom the company is oriented. In recent decades the customer service takes increasingly important in organizations and is regarded as a differentiating factor that allows to develop competitive advantages.

With knowledge of the amount of resources needed in systems, it is essential to manage the ranks to ensure among other things, the quality and welfare as an organization must provide its customers and employees with the aim of generating productivity, maximize profits and minimize costs. That's at that point where queuing theory comes into play a leading role in analyzing systems, performance measurement and quantification of the costs associated with their operation. Its use is also necessary due to uncertainty in the systems, which potentiates even more by involving humans conducted operations there. Human behavior causes the state of a system varies from not very predictable, but despite this, it is possible to govern this behavior by statistical laws, exactly or approximately.

In addition to this, there arises a difficulty that lies in the mathematical complexity of the above suggests, for analyzing systems must reflect the actual situation and the daily operation of organizations composed of multiple points of care and different configurations. Up to 70% of mathematical programming problems in the corporate sector can be represented as models related to networks, making it necessary to develop tools that extend the applicability of queuing theory to these business areas. By analyzing systems modeled networks it is possible to have a broader view inside the system and take into account the synergy between subsystems.

Therefore, it is in their tails networks and where should focus considerable efforts. More however, must go beyond the same utility queuing theory, not only measure performance but to seek improvements using optimization tools for calculating efficient measures of resources, costs, profits, times and other proprietary indicators of each organization. a powerful, flexible and intuitive tool that is extremely useful when it is difficult or impossible resolution analytical or required numerical model. This hybrid will allow analyze complex systems where: In addition, the use of another tool such as simulation analysis is necessary different arrival rates and service, and changes in the number of servers and system resources available are presented.

The foregoing demonstrates the need for organizations to analyze, understand and explore their care systems where waiting lines are generated, in order to diminish the impact they have on their productivity. The queuing networks and broader vision systems, along with the closest approach of non-exponential distributions and new solution methodologies are variations that make this study is very useful to the service sector, closer and closer to their real situation. In addition, the realization of this study from the multidisciplinary perspective that provides industrial engineering, makes solutions encompassing technological, organizational, commercial and economic.

Further benefits perceived by organizations, this development will be a research contribution to queuing theory, to work, analyze and understand complex network systems that can be used in a number of business applications, given a variable and a standardized solution methodology.

5.4. THE OBJECTIVES

5.4.1. General Purpose

Analyze the behavior and performance of queuing networks in service processes for minimizing the costs associated with waiting and residence times in the system efficiently allocate human resources in organizations such structures.

5.4.2. Specific Objectives

Define a queuing network on which it is possible to make modifications of certain factors of the waiting lines, as arrival rates service, probability distributions of inter-arrival times, amount of available servers and workload among others.

Determine via simulation performance possible systems that can be generated in the network queue to combine treatments defined in experiment designed and evaluate how the variation of these parameters in the system impacts efficiency.

Develop a model of multi-objective optimization to minimize waiting times on the network and costs associated with that expected to efficiently allocate resources by applying heuristic algorithms.

Build a computational tool oriented service sector organizations to determine system performance based on the results of the simulation and assign the network resource from the solution of the optimization model.

5.5. THEORETICAL FRAMEWORK

5.5.1. Time Series

The time series are a methodical registration fixed time intervals of the characteristics of a variable, or its numerical observation. They are used to describe and analyze phenomena for over time. Time series are ultimately stochastic processes, but with the restriction that are indexed by time and the cuts are made at fixed intervals. Then we define a stochastic process.

5.5.2. Stochastic Processes

A stochastic process is a family of random variables associated with an index set of real numbers: i.e., each element of the set index number, corresponds one and only one random variable.

Be T the set of real numbers index. We define $Z(\tau)$ as the random variable corresponding to element τ from T (that is to say, τ It's in T). We define as stochastic process family (or the whole) of random variables $\{\ Z(\tau),\ \tau \in T\ \}$

The discrete time series, ($\tau_1, \tau_2, \tau_3 \ldots, \tau_n$) it is observations of a variable in time (1,2...n). The respective stochastic process is: $\{Z(\tau_1), Z(\tau_2), Z(\tau_3) \ldots . Z(\tau_n)\}$. That is, a family (or group) of random variables. Thereafter the name of the variable (t) and its observed value will be denoted by ***Zt***.

A time series observed is simply a realization of a stochastic process there will always be a probabilistic element in recorded and observed observations.

The behavior of a random variable (Z) It described by its density function f (Z). The behavior of two random variables Z_1, Z_2, It is described by the joint density function f Z_1, Z_2 And whether they are independent random variables:

$$f(Z_1, Z_2) = f(Z_1) \cdot f(Z_2)$$

In the time series analysis of the assumption that the observations not come from independent random variables it is established: it is assumed that there are a correlation structure between the observations; so it is not easy to obtain the joint probability density function.

5.5.3. Advantages of Using Stochastics to Time Series

- Flexibility to represent a large number of phenomena with one general class of models.
- Easy and accurate forecasts.
- Generalizing analysis methods individual groups of variables.

Now we analyze the deterministic processes. To do this, we will use certain tools of difference equations.

"Temporarily" we ignore randomness. Then, as will a comparison between equilibrium and stationary.

5.5.4. Difference Equations:

The equation:

$$Z_t = a_0 + a_1 Z_{t-1}$$

can be written as: $(1 - a_1 B)Z_t = a_0$

The general solution of this equation is:

$$Z_t = [a_0 / (1 - a_1)] + s a_1^t$$

It is easy to prove that the solution of the differential equation 1 is given by equation 2:

Equation 2 by applying to the operator; $1 - a_1 B$ Equation (1) is obtained.

The particular solution requires additional information to calculate the value of (*s*). More generally, knowing the initial condition $\mathbf{Z_0}$, We would have to $Z_0 = [a_0/(1-a_1)] + s$ And the solution of the difference equation would be:

$\boldsymbol{Z_t = [a_0/(1-a_1)] + [Z_0 - (a_0/(1-a_1))]a_1^t}$ if we assume that $|\boldsymbol{a_1}| < 1$, when $\boldsymbol{t \to \infty}$; $\boldsymbol{a_1^t}$ tends to zero and $(\boldsymbol{Z_t})$ tends to: $\boldsymbol{Z_t = a_0/(1-a_1)}$

Iterative or method "pedestrian," also solve an equation difference.

$$\boldsymbol{Z_1 = a_0 + a_1 Z_0}$$

$$\boldsymbol{Z_2 = a_0 + a_1(a_0 + a_1 Z_0) = a_0(1 + a_1) + a_1^2 Z_0}$$

$$\boldsymbol{Z_3 = a_0 + a_1 Z_2 = a_0 + a_0(a_1 + a_1^2) + a_1^3 Z_0}$$

Difference equations of second order:

$$\boldsymbol{Z_t = a_0 + a_1 Z_{t-1} + a_2 Z_{t-2}}$$

Time Series Types

5.5.5.1. Time Series That Are Stationary

Once we have a defined function, we can define the stationarity of a stochastic process, in this case a series of time (more specifically we refer to weak seasonality). In general, when we talk about stationarity we mean weak stationarity.

Stationarity and stationarity.-time series $\{X_t, t \in Z\}$, where Z is the set of integers is stationary if and only if:

i) $E[X_t^2] < \infty$, $\forall t \in Z$

ii) $E[X_t] = \mu$, $\forall t \in Z$

iii) $Cov[X_s, X_t] = Cov[X_{s+h}, X_{t+h}]$, $\forall s,t,h \in Z$.

In short a time series is stationary its variance remains finite, its first moment is constant over time and that the covariance between two variables in the series only depend on the span between these and not on its location in time (i.e., the covariance between X_s Y X_t Only depended of $(s-t)$, And not from s or t.

5.5.5.2. Strict Stationarity

It is said that the time series $\{X_t, t \in Z\}$ is strictly stationary if the joint distribution of $(X_{t_1}.X_{t_2},X_{t_3},...,X_{t_4})$ It is the same as that of $(X_{t_1+h}.X_{t_2+h},X_{t_3+h},...,X_{t_4+h})$.

It should be noted that strict stationarity not need that X_t maintain their finite variance therefore strict stationarity not necessarily imply weak stationarity. It is also worth noting that all non-linear function of a strictly stationary process remains strictly stationary, however not the case with non-linear functions of weak stationary processes. For example the square of a weak stationary process may not have a finite variance.

The properties of the stationary and non-stationary processes are very different, requiring different methods to make inferences. A process for nonstationary empirical, simple and useful to recognize between a stationary process studies and one is plotting data. If a series does not seem to have a constant mean or variance, then it is very likely not stationary.

5.5.6. Stationarity of the Time Series

In practice many series are not stationary; but his first and second differences. The purpose of differentiating one series is stationary back to the differential of said series. However, it should be remembered that if differences stationary series which are already taken; these differences will also be stationary. Then we can shop on differentiation of the series; which entails identifying problems regarding that model that best represents the process that follows the series and its variance increases.

A time series is seasonal when and its tendency and long term cycle, shows fluctuations periodically repeated. There may be similarity of behavior for observations of the same month; for example, selling toys in "months of December." There may also be a pattern of periodic behavior with less than one year. For example, "every six months" from June. Observations of "months of June" and "months of December" will be similar in their behavior, in addition to a similar behavior of the observations of the "months of December" to each other, and "months of June" to each other.

It is generally desirable to consider a period where we expect to see E seasonal behavior of the variable.

5.5.7. Ergodic Processes

The law of large numbers states that if the Kolmogorov X_i s are independents and are all identically distributed with mean μ and variance σ^2 then we have the following limit:

$$\bar{X} = n^{-1}\sum_{i=1}^{n} X_i \to \mu$$
.

Time series instead of having an average of a set of individuals in a given time, we will have an average over time the same individual. To better understand this difference first consider a set of students course, average whole course of a final exam was first taken. Then the average school career of one student was taken. They certainly are different sizes. In the first case the law of large numbers states that the average course to converge μ distribution. To make an application of this law, the time series is necessary that they are ergodic and stationary.

To define what is an ergodic process, we must first define what the temporary means of a process. Yes $Y(t)$ It is a stochastic process, then its time average, $\langle Y(t)\rangle$, Defined as follows:

$$\langle Y(t)\rangle = \lim_{k\to\infty}\frac{1}{2T}\int_{-T}^{T} Y(t)\cdot d$$

Now if we can define what is an ergodic process $X(t)$ It is an ergodic process if and only if all its statistical averages family can be exchanged for their corresponding half-time. That is a simple temporal embodiment is representative of the whole process. A process is ergodic implies that it is stationary, but not the other way around.

Therefore: If the time series X_t It is stationary and ergodic with $E[X_t] = \mu$ Then the average time series converges to μ, that is to say:

$$\bar{X}_t = t^{-1}\sum_{i=1}^{n} X_i \to \mu$$

5.5.8. Linear Stochastic Processes

Shocks occur independent random variable. These collisions generate time series for certain other variables, whose successive values can be highly dependent: this is the great idea introduced Yule (1927) and Box and Jenkins (1970).

(a_t) It is the variable subject to independent random shocks. However, the values that are generated for Z_t related to a polynomial lags a_t, Are highly dependent with successive values of Z.

For example: at, the independent variable subject to random shocks, it may be "the amount of rain," which tends to affect "maize crops." It is possible that the corn crop at time t, denoted by Z_t It is affected by the amount of rain the year t, that of a past period and Ancestor period; so the current harvest and next year will be related: both dependent on the current rains and last year.

$$Z_t = \mu + a_t - \Psi_1 a_{t-1} - \Psi_2 a_{t-2} - \dots = \mu + [1 - \Psi_1 B - \Psi_2 B^2 - \Psi_3 B^3 + \dots] a_t$$

$$Z_t = \mu + \Psi(B) \cdot a_t$$

The polynomial delay for a_t "Turns," through a connection, the stochastic process $\{a_t\}$ in the process $\{Z_t\}$.

5.5.9. Features of a Series of Time

To make inferences about these variables, it is often useful to "break" the time series for its main components:

- Trend and cycle: represents the long-term movement of the series.
- Seasonality: represents effects of phenomena that occur periodically or reproduce (weekend, Decembers, Fridays, etc.).
- Irregularity: unpredictable or random movements.

Cycle and seasonal trend and make up the deterministic part of the series, while the irregularity represents non-deterministic or stochastic part of the series.

Many users of information limited to seasonally adjusted series Stochastic (partly by widespread methods of seasonal adjustment), without attempting a more complete statistical analysis.

5.6. TIME SERIES MODELS

5.6.1. Operators and Polynomials

Delay polynomials are very useful because they allow to represent concisely and simpler models that are very valuable (but seem complex).

Delay operator or "backward" Applicable to Zt. It indicates that the variable should be delayed period: that is to say, $BZ\ t\ Z\ =\ t-1,$

as well, $B2Z\ t\ =\ B\ [BZ\ t]\ =\ B\ [Z\ t-1]\ =\ Z\ t-2,$

and in general,

$$Bk\ Z\ t\ =\ Z\ tk.$$

difference operator ∇, applicable to Z_t. It indicates that the difference should get Z_t and its lagged value:

$$\nabla Z_t = Z_t - Z_{t-1} = (1-B)Z_t$$

$$\nabla^2 Z_t = \nabla(Z_t - Z_{t-1}) = (Z_t - Z_{t-1}) - (Z_{t-1} - Z_{t-2})$$

Polynomials formed by observations present and past weighted:

$$G(B)Z_t = Z_t - g_1 Z_{t-1} - g_2 Z_{t-2} - \ldots g_k Z_{t-k} = Z_t - \Sigma g_j Z_{t-j}$$

Rational polynomials

$$G(B) = A(B)/C(B)$$

A (B) = 1-Σj B j; C (B) = 1-Σ B j cj

Example: prove that the next operator (polynomial delay) is "rational": Clearly in this case G (B) is a rational polynomial delay, since it is equivalent to the ratio:

G (B) = A (B)/C (B) = 1/(1-g B). Where: A (B) = 1; C (B) = (1-g B).

Models Autoregressive (AR): Deterministic. They are of the type: $A(B)Z_t$ = constant Stochastic. a random variable is introduced: $A(B)Z_t =$ constant $+a_t$

Where $\{a_t\}$ It is a white noise process. $A(B)$ is a lag polynomial of the form: $(1-\phi_1 B - \phi_2 B^2 - \phi_3 B^3 + \ldots..\phi_p B^p)$. Then the stochastic model can be written as:

$$\left[1-\phi_1 B-\phi_2 B^2-\phi_3 B^3-\phi_4 B^4+....\phi_p B^p\right]Z_t = \text{constant} + a_t$$

Yes: $E(Z_t)=\mu$, for all t It implies that the constant is equal to $(1-\phi_1-\phi_2...\phi_p)\mu$. And that the stochastic process can be written as:

$$\phi(B)\tilde{Z}=\left[1-\phi_1 B-\phi_2 B^2-...\phi_p B^p\right]\tilde{Z}_t=a_t,\ \tilde{Z}=(Z_t-\mu).$$

Autoregressive process is because it can also be written as:

$$Z_t=(1-\phi_1-\phi_2...-\phi_p)\mu+\phi_1 Z_{t-1}+\phi_2 Z_{t-2}+\phi_p Z_{t-p}+a_t$$

It is important to determine if the stochastic process is stationary; I mean, yes (Z_t) It tends to be localized (most often) in the vicinity of its average value.

We have:

$\phi(B)\tilde{Z}=a_t$. The process will be stationary if for lag polynomial, ϕ (B) equal to:

$$\phi(B)=(1-g_1 B)(1-g_2 B)....(1-g_p B),\ g_i<1,\ \text{for } i=1,2,3...p$$

Representation of processes (or models) autoregressive (AR):

$$(Z_t-\mu)=\phi_1(Z_t-\mu)+\phi_2(Z_{t-1}-\mu)+...\phi_n(Z_{t-k}-\mu)+a_t$$

The AR model is of order (1,2,3,... p), according to the number (1,2,3,.. p) of lags lag polynomial operator θ (B) perform.

Autoregressive model of order one. AR (1):

$\tilde{Z}_t-\phi\tilde{Z}_{t-1}=a_t$, Which generates a time series known as Markov series. The process can be re-written as: $(1-\phi B)\tilde{Z}_t=a_t$. The root of the equation: $1-\phi X=0$, We must be outside the unit circle: i.e., $\phi<1$.

Model AR (2):

$$\left[1-\phi_1 B-\phi_2 B^2\right]\tilde{Z}_t=a_t.$$
$$(\tilde{Z}_t=Z_t-\mu)$$

Model AR (p):

$$\left[1-\phi_1 B-\phi_2 B^2-\phi_3 B^3-.....\phi_p B^p\right]\tilde{Z}_t=a_t.$$

$$\tilde{Z}_t = \phi_1 \tilde{Z}_{t-1} + \phi_2 \tilde{Z}_{t-2} + \phi_3 \tilde{Z}_{t-3} \ldots\ldots + \phi_p \tilde{Z}_{t-p} + a_t$$

Models moving averages (MA).

Introduced by Yule (1926) and Slutsky (1927). They are stochastics $\{Z_t\}$ whose values can be dependent on each other, because they correspond to a finite weighted sum of independent random shocks $\{a_t\}$.

$$Zt = (1 - \theta_1 B - \theta_2 B2 - \theta_3 B3 - \theta_4 B4 . - \ldots\ldots\ldots \theta_q Bq)$$

The term "moving average" is not strictly correct, because θ_i It can be negative and $\Sigma \theta_i \neq 1$. θ_i is less than infinity and considering a finite number of summands, the process is stationary.

Moving averages representation by polynomials delays:

$$Z_t = \mu + a_t + \theta_1 a_{t-1} + \theta_2 a_{t-2} + \ldots + \theta_n a_n$$

$$(Z_t - \mu) = (1 + \theta_1 B + \theta_2 B^2 + \theta_3 B^3 + \ldots + \theta_n B^n) a_t$$

Where $\{a_t\}$ It is a sequence of random variables with certain characteristics.

The compact representation model moving average (MA) will be:

$Z_t - \mu = \theta(B) a_t$. This model is known as the moving average model, or model MA for its acronym in English. According to the number of lags that the polynomial operator θ (B) perform (1,2,3,... q), the MA model will reach an order (1,2,3,... q).

Model MA (1):

$\tilde{Z}_t =$ (1- θB) at t = a θ t-1

AND $(\tilde{Z}_t)$ = 0; VAR $\tilde{Z}_t$ = (1- θ^2) VAR (a) = (1- θ^2) σ_a^2.

The auto-covariance between $\tilde{Z}_t$ Y $\tilde{Z}_{t-k}$ be:

E [(a t – θ t-1) (a t – θ tk-1) = $-\theta\sigma_a^2$ For K = 1 and zero for $K \geq 2$;

so the autocorrelation coefficient between Z_t Y Z_{t-k} be:

ρ_k = Auto-covariance ($\tilde{Z}_t ; \tilde{Z}_{t-k}$)/VAR (a) = -$\theta$(1-$\theta^2$) For K = 1; zero $K \geq 2$.

The MA (1) process "does not remember" what happened beyond the previous period; i.e., have a limited memory period. I feel that $| \rho_k | \leq 0.5$. Autoregressive stationary process can be represented by models of moving averages (provided ϕ is less than one); similarly the MA (1) models may be represented as an autoregressive model if
$| \theta | \leq 1$.

In general when one o'clock (q) process can be expressed by an AR (p) is said that this process is invertible.

Model MA (2):

$\tilde{Z}_t =$ (1- θB - θ_2 B2) at

$(\tilde{Z}_t = Z_t - \mu)$

AND $(\tilde{Z}_t)$ = 0; VAR $(\tilde{Z}_t)$ = (1 + $\theta + \theta^2$) σ_a^2.

$E(\tilde{Z}_t) = 0$ and an autocorrelation coefficient series ρ_k > 0 for K = 1,2,3... q, $k \geq q+1$; which it tells us that the MA (q) process has "limited memory" periods. Autoregressive and moving average: stochastic process that follows the random variable Zt, whose deviation from its expected value μ we denote by:

$\tilde{Z}$ t = Z t – μ

The model is expressed as follows:

1) $\phi(B)Z_t = \theta(B)a_t$,

Where ϕ (B), θ (B) are operators lags order p and q respectively, at {} is a random variable with white noise process (zero mean and finite variance).

An alternate way of writing the process followed by the variable $\tilde{Z}$ t would:

$$(1+\phi_1 B+\phi_2 B^2+...+\phi_p B^p)Z_t = (1+\theta_1 B+\theta_2 B^2+...+\theta_q B^q)a_t \quad (1.1)$$

O well:

$$Z_t+\phi_1 Z_{t-1}+\phi_2 Z_{t-2}+...+\phi_p Z_{t-p} = a_t+\theta_1 a_{t-1}+\theta_2 a_{t-2}+...+\theta_q a_{t-q} \quad (1.2)$$

The ARMA (p, q) is a generalization of the AR and MA models, combining both classes of models. Such generalization arises noted that the time series simultaneously present process characteristics AR and MA. In addition, the principle of parsimony suggests building models that include the smallest possible number of parameters.

It is expected that not all be stationary time series, of course under which the ARMA model is built. However, we know that for almost any non-stationary series, the first, second or third difference if the series is stationary. Under these conditions Yaglom (1955), found that if the original process $\{\tilde{Z}_t\}$ Suffers from stationarity caused by a nondeterministic polynomial trend (which is called inhomogeneous stationarity) is possible to construct a stationary process Wt {}, such that:

2) $W_t = \nabla^d \tilde{Z}_t$. For all t. For this new series it is possible to obtain an ARMA model:

$\phi(B) W_t = \theta(B) a_t$ equivalent to consider the ARIMA model: $\phi(B)\nabla^d \tilde{Z}_t = \theta(B) a_t$

In the ARIMA model the term "integration" comes from Z t is the sum of an infinite number of current and past values of Wt. Consider equation 2 for d = 1. The value of Zt can be obtained by multiplying both sides of this equation by the operator∇-1, We get:

$z_t = \nabla^{-1} W_t = (1-B)^{-1} W_t = W_t + W_{t-1} + W_2 + W_{t\,t-3} + \ldots$, a sum of an infinite number of terms.

$$ARIMA\ (p, d, q)$$

Since in many cases the stochastic process followed by $[Z_t - \mu] = \tilde{Z}_t$ It is not stationarity; but if your differential first, second, third… Nth order, we can make a generalization ARMA model to reach what is known as ARIMA model.

Finally we have:

$\phi(B)\,[\nabla^d (Z_t - \mu)] = \phi(B)\,\nabla^d\,\tilde{Z}_t = \theta\,B_t.$

Which is the integrated model called autoregressive and moving average, or ARIMA model for its acronym in English (Autoregressive, integrated, moving average). The ARIMA model is more precisely described as: ARIMA (p, d, q). Where p is the number of lags lag operator polynomial ϕ (B) takes, d is the number of differentiations on $\tilde{Z}_t$ the operator ∇d performs q is the number of lags lag polynomial operator θ (B) takes.

An ARIMA (p, d, q) indicates that the model has an autoregressive polynomial of order p, a differentiation in the study variable $\tilde{Z}_t$ order d, and a polynomial of order q moving averages.

Seasonal multiplicative model $(ARIMA\ (p, d, q)\ X\ (P, D, Q)\ E)$: To incorporate seasonal and stochastic effects no seasonal subject observed values of a certain characteristic of the population, or time series, Box and Jenkins (1970) suggested a general model of the type:

$$\Phi(BE)\nabla_E^D (Z_t - \mu) = \Theta(BE)\alpha_t$$

Where the variables $\{\alpha_t\}$ Are assumed not white noise but generated by a process ARIMA (p, d, q), i.e.:

$$\phi(B)\nabla d\ \alpha_t = \theta(B)\, a_t$$

With (at) white noise process. Of the latter two expressions seasonal multiplicative model is obtained:

$$\phi(B)\Phi(BE)\nabla_E^D (Z_t - \mu) = \theta(B)\Theta(BE)\, a_t$$

Which is denoted by: ARIMA (p, d, q) X (P, D, Q) E.

As expected, the more complex the model corresponds to a structure more complex autocorrelation.

The ARIMA model for seasonal multiplicative series with monthly observations allowed;

1) Consider the relationship that may exist between observations contiguous months within year.
2) Consider the relationship that may exist between years for observations of the same month.

That is "captures" simultaneously, seasonal effects and tendency of "multiplicative" process or "self-reinforcing" so such effects.

5.7. STOCHASTIC PROCESSES

A stochastic process is defined as an indexed collection of random variables $\{X_t\}$, where the index t takes values in a set T die. Often T is taken as the set of non-negative integers and X_t represents a measurable characteristic of interest at time t. For example, X_t may represent the number of people in a system at the end of period t. Interest of stochastics is to describe the behavior of a system in operation for some periods. To know their behavior, there are many applications and is also used as sustenance for many probabilistic theories one is queuing theory.

The detection of clusters (clusters) is of great interest in many areas. In epidemiology, for example, it is important to know whether certain diseases occur more frequently in certain geographic areas, resulting in an abnormal case grouping. Resource allocation by the local police to the various city districts should be made taking into account the possible existence of clusters of increased crime. The unexpected and unexplained accumulation of accidents in certain areas of a city or highway, demands the attention of the traffic authorities. All these examples, and many more could be cited, previously required to check that indeed in the geographical area under study observed the phenomenon occurs more often than we would expect.

A simple way for the theoretical concepts required is that which we present below, although it can be more sophisticated and efficient methods to address the problem. Suppose that to facilitate the incidence and location of the event that interests us. This type of quotients called odds against and tells us how many times it is more likely that an event does not happen against occurs. If (1.1) is clearly greater than 1, we will be inclined to reject the hypothesis of the existence of a cluster in the observed data.

For the calculation of (1.1) use the formula of Bayes requiring knowing P (B), P (AJB) and P (AJB^C). Let's see how we can know them. The pro- likelihood of there being a cluster depend on the phenomenon under study and our knowledge of it help us to assign a priori value P (B). If we believe that a cluster is very unlikely, assign a very small value ~ not, for example P (B) = 10¡6. The other two are, respectively, the odds of having observed the event 11 times in the shaded area according to admit it or not the existence of a cluster. To calculate this we note that in each cell or event does not occur independently of the others and that makes all of them with equal probability. That is, the occurrence of the event in each cell can be likened to a Bernoulli test and therefore the total occurrences in 145 cells will be a binomial random variable.

The problem of serving customers who arrive in a queue, is vital in many activities. It is to reconcile an efficient service to the customer, minimizing your waiting time with a rational use of available resources. Obviously operate a large number of posts of attention is a solution, but certainly not the best for the company.

Imagine a simple situation and see how to cope by resorting to a well-known probability distribution, the Poisson distribution. Suppose for this rush hour supermarket, between 7 and 8 pm when people take back home from work to do some shopping imperative, which are not very numerous. The manager of the supermarket opens every day at that time a quick cash, no more than 10 articles, but has observed that lately accumulate in her customers and, what is worse for your business, clearly show his dissatisfaction bemoaning the lack of service. To remedy the situation has decided to resort to an expert, it is assumed that probability for advice on how many additional boxes must be opened.

The experience accumulated over time lets we know that the average duration of care to customers of the fast tail is 1 minute, and what we want is that in 95% of cases there is more than one person waiting to be served. Given the minute it takes to be served, ideally at most reach 2 people per minute to the box.

The first thing the expert was to observe the total number of people who were treated at the only quick cash available between 7 and 8 pm. Logical observation made over several days, from Tuesday to Friday, and obtained results 68, 70, 59 and 66 clients respectively. That is, on average about 70 customers when or 1; 167 per minute. Moreover, the expert interpret, \... that in 95% of cases there is more than one person waiting to be served "in terms of probability, namely that P (N · 2) = 0; 95, where N is the variable representing the number of people in the checkout line. the characteristics of the problem no doubt offered expert on the probabilistic behavior of N, it was a Poisson random variable.

Recall that a Poisson variable takes non-negative integer values, N = 0; 1; 2; 3; … and its function is of the form

$$FN\ (K) = P\ (N = k) = exp\ (n)\ k!\ :$$

The problem for the expert was to know the value of the parameter, but that made his remarks because depends on the characteristics of the phenomenon and represents the average number of occurrences of the event under consideration per unit time.

The Poisson distribution and the exponential distribution arise naturally in the so-called Poisson process, which we shall discuss in detail in the chapter on stochastic processes. PA effects we are now interested enough to do a simple description. A Poisson process arises when we deal with the occurrence of an event over time: The calls that reach a PABX telephone, radioactive disintegrations that reach a Geiger counter, customers who reach a point of care, accident in a nuclear plant,.... to study this phenomenon certain simplifying assumptions are made,

1. the different occurrences of the event are independent of each other,
2. the probability of two or more occurrences of the event in a small range ~ no time is practically nil, and
3. if I1 and I2 two time slots are such that I1 / I2 = k. The random variables N1 and N2, Designating the number of occurrences in each are independent.

With these hypotheses, it can be shown that the number of occurrences in any interval of length t follows a Poisson distribution of parameter T that when determining the distribution of Nt all that matters is the length of the interval and not where it is located, this property is called stationarity.

Discrete variables with Poisson distribution denoting the number of occurrences of the event in the interval of length t, and Xt, Continuous variables that denote the time between two consecutive occurrences of the event time, the ith and (i – 1)-th. Given the independence between occurrences, they are independent and, of course, all have the same distribution. We obtain the common distribution function. Remembering that

$$Fi(T) = P(Xi \cdot t) = 1P(Xi > t);$$

which it is the density function with exponential parameter
Poisson process could also have been defined from the times between consecutive occurrences of the event. If we posit hypothetical independence of such times as common distribution Exp(,) then distributed as Nt?

5.8. ENTROPY AS A MEASURE OF INFORMATION

By carrying out the experiment bound to variable X whose support is DX = {x1; x2; :::; xk}, The result will be X = xi. A party interested in this result and to know makes a series of questions that support only response yes or no. What will be the average number of questions there will pose to know the result? There is a minimum for this average? Before responding, and to establish the relationship between the response and HX, Let's look at an example to help understand the problem we have raised.

Example. A 32 urn containing numbered balls from 1 to 8 being its composition as shown in Table 2.1. It is extracted at random and we want to know what strategy to follow to minimize the number of questions needed to know the number extracted.

digit	1	2	3	4	5	6	7	8
number of balls	8	8	4	4	2	2	2	2
P (Ball = i)	1/4	1/4	1/8	1/8	1/16	1/16	1/161/16	

Table 2.1: Composition of the urn

Since the numbers that appear in more balls are more likely, a reasonable strategy is to ask for the numbers in descending order of probability

If we represent by N1 and N2 the number of questions required for each strategy to know the number of the ball drawn, their values depend on that number and can be obtained.

ball drawn	1	2	3	4	5	6	7	8
value of N1	1	2	3	4	5	6	7	7
value of N2	2	2	3	3	4	4	4	4
P (Ball = i)	1/4	1/4	1/8	1/8	1/16	1/16	1/16	1/16

The problem of designing a strategy ~ nar questions with dichotomous response to identify exactly the value of the variable X = f-number that shows the extraidag ball is the same that occurs when we encode the output of an information source. Indeed, the sequence of responses that leads to the identification of the value of X can be likened to a sequence of 0's and 1's, according to the responses have been negative or positive, respectively. It is ultimately a binary code and the problem of finding the best strategy questions is equivalent to finding the shortest binary code.

Two fundamental results of information theory allow us to establish the important role of the concept of entropy. The enunciate without proof.

1. The average length of any binary code cannot be less than the value in bits of entropy.
2. If the values of the function on the intensity of X are powers of 2, there is a strategy (Codification) whose average value equals the entropy. As with the second strategy the previous example.

As a result of these two results we can say that the entropy of a random variable X is the lowest average number of bits needed to identify its value.

5.9. MARKOV CHAINS IN CONTINUOUS TIME: LIABILITY OF A MULTIPROCESSOR

They have a computer with two independent processors and we want to model the behavior of the system over time. It is a system with three states:

sl = 0, it is indicating that both processors do not work.

sl = 1, indicating that only one of the processors works.

sl = 2, it is indicating that both processors work.

The probabilistic model describing timeouts, either a failure or a repair, is the exponential model. Assume therefore that the failure time TF "Exp (,) and repair time Tr "Exp ('), and both are independent.

The process Xt; t¸ 0 designates the state of the system at time t. It is a Markov chain continues over time and homogeneous. To check this we get the transition times for each state, we use the exponential property of memory.

Transition 1. A transition of this type occurs when both processors are out of service and one of them is repaired. If T01 is the transition time corresponding to Trl and T r2 times repair processors, T01 will coincide with the time it is first repaired, then

$$T01 = Min\,(Tr1;\, Tr2\,);$$

and here

$$P\,(T01 > t) = P\,(Min\,(Tr1;\, Tr2\,) > T$$
$$= P\,(Tr1 > T;\, Tr2 > t)¶$$

5.10. QUEUING

Many service industries have a queuing system in which products or customers arrive at a station and waiting in line, get some service and then leave the system. This type of phenomenon occurs when users reach a certain service more quickly to this is capable of dispatching, and for that reason people or objects should expect to be accumulated.

Queuing theory is the study of that expected in various forms, using queuing models systems lines expected to arise in practice are represented. The formulas for each model indicate which should be performed corresponding system and indicate the expected amount of time and people in a queue, in a range of circumstances.

The random nature of phenomena queue evidently implies that the main weapon of this analysis is the calculation of probabilities. The set of formulations and relationships using data at a certain probability distribution is a mathematical model is called stochastic process in which one or more magnitudes vary randomly according to the basic structure of queuing models.

A waiting line is constituted by a customer who requires a service that is provided by a server in a given period. Randomly customers enter the system and form one or more queues to be served. If the server is idle, according to certain rules preset by the name of service discipline, service to queue items in a given period of time is provided, called service time and then leave the system. Waiting lines can be classified according to [53] using (see Figure 5.1):

- The number of customers who can wait in line and population;
- Elapsed interval between a client and another time;
- Service time;
- The queue discipline. (FIFO, LIFO, etc.)'
- The number of servers;
- The structure of service stations; and
- System stability.

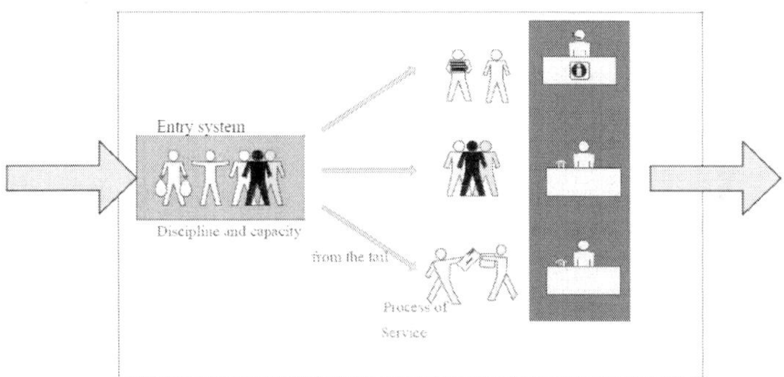

Figure 5.1. Structure of queuing systems.

5.10.1.Classification System

The classification system is the best known and Lee Kendall in 1953 developed in order to standardize similar models and thus analyze alike. This is beneficial not only in terms of the notation, but opens the possibility of associating certain types of formulas and calculated systems with similar characteristics in case of exponential distributions.

- Probability distribution of time between arrivals of entities;
- Probability distribution service time;
- Number of servers;
- Discipline care;
- Maximum number of entities supporting the system in a period t and
- Population.

For example, an M/M/1 system means a time of arrival and attention exponential distribution and a single server serving said queue; and other features are specified it is assumed discipline is general and the other infinite parameters. In the case of a G/M/S we would be talking of arrival with a general distribution, exponential attention and S servers. Due to the complex management of non-exponential distributions in queues to find the values to the unknowns is necessary simulation.

5.10.2.Performance Measures

In the analytical review of the queues there are two types of performance measures that are most important and used, the two measures are usually expressed in terms of their expected values are:

$$L = expected\ number\ of\ customers\ in\ the\ system$$
$$W = expected\ waiting\ time\ in\ the\ system$$

Now, if not account people who are seen but only the tail has the same two formulas but with subscript q (Lq, Wq) 56

5.11. QUEUE NETWORKS

So far we have taken into account any more queuing systems that have a service facility with one or more servers, but queuing systems found in studies of operations research, actually sometimes queuing networks, i.e., networks of service facilities where customers request a service of some or all. (Figure 5.2).

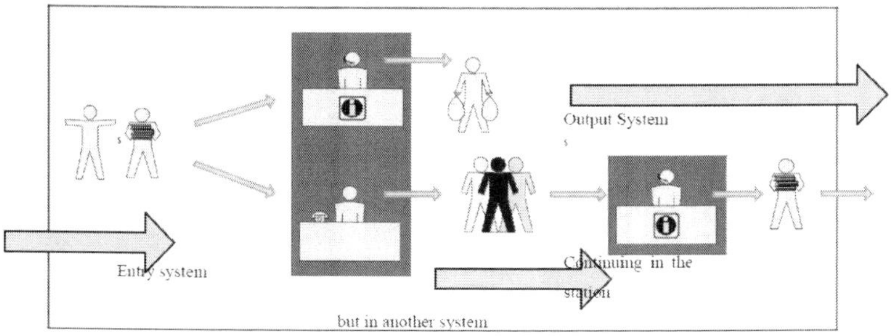

Figure 5.2. Networks Queue.

A queuing network is a set of systems connected to each other queues. In such structures customers arrive a node or point of service and wait their turn to be served as in a conventional tails; the difference is instantly leaving a node itself may or may not go to another to receive an additional service or continue a service in a different unit, this happens indefinitely until all service requirements remain.

5.11.1. Classification of Queuing Networks

Open networks: Customers receive input from one or more external populations and have outputs from outward. They are used in service centers and more applications. Its two subtypes are:

- acyclic: A job can never return to the same queue.
- cyclical: There are loops in the network

Closed networks: not receive inputs from external populations or have outlets abroad. Constant number of customers circulating within the red section.

Figure 5.3. Classification of queuing networks

5.11.2. Jackson Networks

Jackson network is a system where installation facilities i (i = 1,2,..., m) satisfy the following:

- A queue of infinite capacity.
- Customers arriving from outside the system according to a Poisson process input parameter a_J
- If a number of servers, with the same exponential distribution with parameter μi for service times.
- A customer leaves the installation i can exit the system, or can go to another system j (j = 1,2,..., m and j ≠ i), with probability . The probability of leaving the system is:

$$q_i = 1 - \sum p_{ij}$$

Jackson networks are so called because it was Jackson who discovered a property that is vital for analysis: Under steady-state conditions, each facility j (j = 1,2,..., m) of a network, It behaves like a queuing system m/m/s independent, with arrival rate m

$$\gamma_j = ai + \sum_{i=1}^{m} \gamma_{ij} p_{ij}$$

5.11.3. Performance Measures

Average number of jobs in the system, which is the sum of the average numbers of jobs in each of the nodes:

$$L_t = \sum_{i=1}^{K} L_i$$

Average time on, W_t system that is the average time it passes a task from entering the network until out of it:

$$W_t = \frac{L_t}{\gamma_t}$$

Property Equivalency: Suppose that a service facility has S servers, a process input parameter Poisson λ, and the same distribution of service time for each server with parameter μ, ρ where <1. Then the steady-state output of this installation service is also a Poisson process medium λ.

5.11.4. Queuing Applications

Queuing theory has had a prominent place in modern analytical techniques, but here the focus is limited to the formulation of a mathematical theory descriptive. So, no concerns directly to queuing theory achieve the goal of operations research: optimal decision making. However, a necessary for this purpose can be achieved with the simulation and optimization input.

5.11.5. Simulation

Simulation is a numerical technique for conducting experiments comprising certain types of mathematical and logical relationships, which are necessary to describe the behavior and structure of complex real-world systems over long periods of time.

Within the wide variety of simulation packages available on the market, the best known and used today they are:

- Simulator is used to model processes, reported and documented. Simulating future system response face to understand complex relationships and identify opportunities for improvement. Operations with dynamic display animated graphics. Analyze how the system works "As is" and under a set of possible alternatives so as to choose the best of certain configuration.

5.11.6. Optimization

It is the use of mathematical models, with the aim of providing knowledge for making decisions. Seeks to improve process performance, to determine how they can maximize or minimize benefits

5.11.7. Types of Optimization

There are several types of optimization used for different situations because of their complexity constitute a problem. Some of these methods are:

- ***Linear programming***: Procedure for solving problems through formulation through linear equations, optimizing the objective function it is restricted to taking by linear equations.
- ***Integer programming:*** Like the linear uses a method or algorithm to solve problems. In these models some or all variables must be whole values.
- ***Dynamic programming***: A general approach to solving problems that need to make decisions in successive stages. Decisions made

at one stage determine the future evolution of the system and affect. Situations where the system will be in the future (called states), and the decisions that will arise in that moment.

- ***Meta-Heuristics:*** It is a method for solving a general problem type, using the parameters given by the user on a generic and abstract procedures in a way that is expected efficient. The meta-heuristics are generally applied to problems that have no specific algorithm or heuristic to give a satisfactory solution; or when it is not possible to implement the optimal method.

Some meta-heuristics greater use and therefore more known are:

- random optimization;
- Local search;
- Best-first search;
- Optimization based on ant colonies;
- Tabu search;
- Genetic algorithms;
- Mathematical algorithms;
- swarm intelligence;
- Stochastic Diffusion Search; and
- extreme optimization.

5.11.8. Queue Networks in Service Centers

Service centers have evolved and, today, have very specific physical, technological and procedural characteristics that go beyond what queuing theory has attempted to model and explain so far. The vast majority of efforts and findings in this theory have focused on the analysis of lines of simple waiting, leaving aside the study of more complex structures such as queuing networks, which represent most successful way centers current service. This chapter describes an experiment which will analyze and determine how variation of specific factors waiting lines impacts the behavior and performance of queuing networks in service processes will be designed.

5.11.9. Definition of the Network Queue

Service processes are generally complicated structures whose composition depends on the arrival or departure flows towards the point of care or to service nodes. Because of this, each focus their convenience has a shaped distribution network to serve customers. Within these structures are the networks type star, bus, cyclic, mixed, ring, tree, and fully connected.

Serial nodes are the simplest type of network and witty; represent a structural basis within all other networks and hence the importance that its use in this study. Using serial networks is the best way to build consecutive processes not limited to particular cases and with the ability to model various structures care by expansion processes, binding or combination. Because of this, it works over a network in series with constant number of cycles between nodes without one or more nodes.

The selection of acyclic serial network simplifies the operation of real systems by splitting the entire care process threads in series. Accordingly, customers will enter through a single point, be cared for at each node according to the capacity of the willing servants and finally exit the system.

In addition, it is assumed that the network will not reach entities massively, there will not be collected for the attention of entities priorities, discipline row will be FIFO, a different queue is created on each node, the number of servers change per node and each node will have a different capacity constrained. (Figure 5.4)

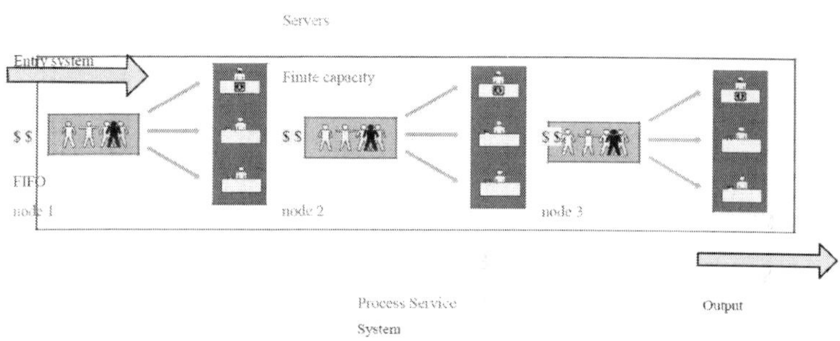

Figure 5.4. Structure of the queuing network defined

5.11.10. Influence Factors

According to David Anderson in his book Quantitative Methods for Business, the behavior of waiting lines is determined by six variables or key features that Kendall and Lee together to identify and classify the various queuing models. The probability distribution of the inter-arrival times and service times, the number of servers, the system capacity, the queue discipline and the population size or source of origin, are part of this notation They allow we to build different models and determine the performance measures system performance by varying characteristics.

5.11.11. Probability Distributions of Inter-Arrival Times

Queuing networks, and generally queuing theory, they have been studied mostly under the exponential probability distribution because of its mathematical ease. Its application became very popular, however, with the computational tools that today there is possible to study much more complex than those at the time Jackson and Burke conceived, shifting the focus to new probability distributions to model time between arrivals. Although in the past the exponential distribution proved very useful in this paper use means a mathematical simplification of what really happens in a system, generating incongruous results in most cases.

The exponential distribution rule out the possibility of inputs to the system by sampling times between arrivals, since it is a dispersed distribution whose mean and standard deviation of arrival times is the same assumption that is fulfilled in very few opportunities in the business environment. In addition to this, the fact of having a single parameter does not allow effectively model the behavior of a system in operation, for the more parameters contains a distribution is more likely that the model can explain the realization.

Because of the above, the use of other probability distributions different to model the exponential inter-arrival times, in order to meet their shortcomings and have representations increasingly approximate reality becomes necessary. Distributions should be sought more versatile, more robust probability, more applicability to the business context of care that is where they are needed.

Recently, the probability distributions obtained by mixing or additions of parameters have shown to be the most appropriate to fulfill this purpose, they are dispersed and have longer that the distribution starting queues, resulting in better fits on data obtained in sampling times.

Within this type of distributions, Hyper-Erlang or Erlang mixture, stands to give more time between arrivals variability, because it has greater number of parameters. Additional parameters, this distribution have sub-phases, each of which has the property of not memory, which provides part of the mathematical simplification of the exponential distribution. Additionally, it offers the possibility to show better dispersion problems queue with less processing time.

Hyper-Erlang distribution resulting from the interrelation of Poisson processes in sequence or phase, making the behavior of system input a more robust and complete process. Modeling network traffic type phase distributions in the past has proven to be a powerful means to facilitate evaluation of the performance of a queuing system and reach conclusions.

One of the biggest advantages of the Hyper-Erlang distribution is that it has proven to be able to approach any positive random variable according to data collected in sampling gives the flexibility to be used in different contexts and provides the accuracy in assembling the bottom behavior. For example, Feitelson & Rudolph presented their work "Job Scheduling Strategies for Parallel Processing" using such distribution for modeling systems where they perform an efficient allocation of work and it is necessary that this allocation possesses an occurrence probability according to a Poisson process. The aim of this study was to reach an efficient allocation of work according to the variability of inputs throughout the process.

On the other hand, this same distribution has been applied to the modeling of mobile communication networks worldwide. Yeoy and Chi identified variables processes arrivals to those networks that Hyper-Erlang distribution modeling better than any other, due to the unpredictable behavior of the cell lines studied. Under this line, they developed a series of studies by Zhang and Hu aimed at analyzing the capacity of a wireless network to support the massive computer connection, while avoiding security problems, sluggishness and saturation. The model developed by Zhang and Hu was based on Erlang distributions Híper to ensure a more accurate modeling of the inputs to the system and perform calculations estimates about the need to expand coverage or wireless network without problems saturation.

Finally, Wang and Zhou proved the advantages possessed hyper Erlang distribution with respect to their peers, as the Weibull and Pareto distribution. In this experiment patterned web traffic, characterized by having high variability in their data, Hiper-Erlang distributions, Weibull and sarong. The main result was the shortest time evidenced processing Híper Erlang because of its simplicity and its closest approach to that seen in a real traffic system internet.

Erlang distribution hyper-is a continuous probability distribution with $\lambda > 0$ parameters, $I > 0$ (number of phases), and $C \geq I$ (shape parameter).

Its density function is:

$$f(x) = \sum_i ai \frac{\gamma_i^C}{\pi(c)} x^{c-1} e^{-\gamma x}$$

where αi is the probability of occurrence of each phase ($\Sigma ai=1, ai\geq 0$)

$$F(X) = 1 - \sum_{i=1}^{L} ai\, e^{-\gamma x} \sum_{j=0}^{Ci-1} \frac{(\gamma_i\ x)^j}{\pi(j+1)}$$

The moment-generating function of a random variable X with distribution hyper- Erlang is:

$$E(X^n) = \sum_{i=1}^{L} ai\, \gamma_i^{-n} \frac{\pi(ci+n)}{\pi(ci)}$$

5.11.12. Distributions of Probability of Inter-Arrival Times to Contrast

According to the above, they are to contrast two probability distributions to model inter-arrival times: the exponential distribution and Hyper-Erlang distribution. In order to compare the results must take equivalent steps including and similar parameters estimated distributions selected probability. This assumes that the lambda parameter exponential and Hyper-Erlang is the same.

The Lambda arrival rate or the system is the expected number of arrivals per unit time is determined from the formula using the system as follows:

$$\rho = \frac{E(X)}{E(t)}$$

For Exponential Distribution

$$\rho = \frac{1/s\mu}{1/\gamma}$$

$$\gamma = \rho s\mu$$

For the hyper-Erlang

$$\rho = \frac{1/s\mu}{lc/\gamma}$$

$$\gamma = \rho cs\mu$$

Where μ is the rate of attention, s the number of servers, c order hyper-

Erlang and I the number of phases. The probability distribution Hyper-Erlang has a shape parameter that should be estimated as well as a number of phases and a probability of occurrence of each of these. To set it a value to these parameters, the expected value of the probability distribution is taken and lambdas are assumed equal for each of the phases, as shown below:

$$E(X\ ) = \sum_{i=1}^{L} ai\,\gamma_i^{-1} \frac{(ci+1-1)!}{(ci-1)!}$$

$$E(X\ ) = ai\gamma^{-1} \frac{(c)!}{(c-1)!} + a_{i+1}\gamma^{-1} \frac{(c)!}{(c-1)!}$$

$$E(X\ ) = (ai\gamma^{-1} + a_{i+1}\gamma^{-1})\ \frac{(c)!}{(c-1)!}$$

$$E(X\ ) = (\frac{1}{\gamma})\ \frac{(c)!}{(c-1)!}$$

$$E(X) = \frac{c}{\gamma}$$

As a result of this procedure we are obtained if the parameters at each stage of hyper-Erlang distribution are equal and C = 1, an exponential distribution is generated; and if C = 1 February Erlang distribution will. This shows, that from at or above 3 values random variables have hyper-Erlang.

Referencing this finding, the missing parameters are set as follows:

C (Erlang distribution order) = 3. From this value guarantees the existence of a hyper-Erlang distribution.

I (number of phases) = 2. This parameter is the interaction between phases and make the phase type property, which allows determining a modeling more accurately and with a wider range sucesos80.

Alpha = 1 / I = ½. Thus the same probability of occurrence is assigned to both phases.

In conclusion, the system performance will be contrasted with arrival times exponential parameter lambda; and arrival times Hyper-Erlang with parameters lambda, C = 3 and two phases with the same probability of occurrence (½).

Table 1. Summary of parameters arrival distributions

Distribution	Exponential	Hyper- Erlang
lambda	$\rho s\mu$	$\rho c s\mu$
Order (form)	1	C = 3
phases	1	I = 2
Probability (phases)	1	$\alpha = ½$

Current systems queues can be very diverse in their shape and size. Because of this, it is not enough to define a serial network to address the stated objective, it is also necessary to determine its length.

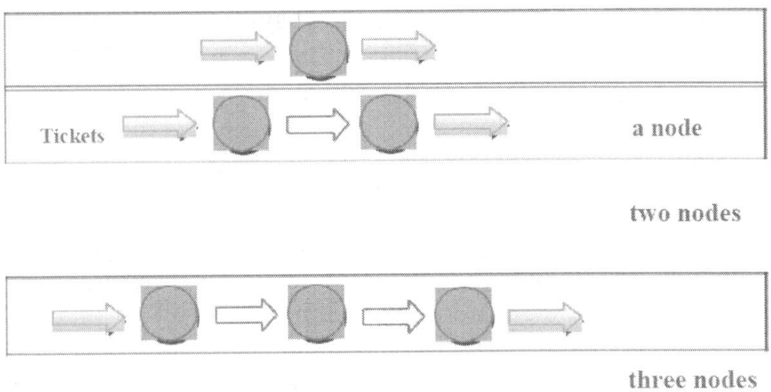

Figure 5. Number of nodes in series

Figure 5 shows modeling networks with the number of nodes that will be contemplated in this study. Accordingly, three different networks that can be a framework to develop a statistical study stochastic queue to be built; and also a means to represent real systems rows.

5.11.14. Number of Servers Per Node

Within each node or point of care there are a number of servers that resolve customer flows and evacuate the ranks. For the evaluation of the number of servers required at each node and its impact on the system, this amount will vary by reference charts and tables showing Hillier and Liberman in the chapter on Queuing Theory Book Introduction to Operations Research. In these graphs, the variation of the average number of customers in queue is shown in an M/M/S with respect to the utilization factor for 1, 2, 3, 5, 7 and 10 servers.

This amount of available node servers vary in each of these, as in many processes there is no homogeneity in the service node to node, i.e., the number of people attend such reception is different from the number server serving later in the same process. Therefore, these amounts will vary in each of the nodes to form combinations of care between each point in series.

5.11.15. Percentage Utilization per Node

Within each node there is a link called system utilization factor that determines how congested this system, what impact it generates the correct operation and steady state operation is achieved.

Rho or utilization factor is the division between the expected time of attention, the expected time between arrivals system and means the expected fraction of time that the servers are occupied; for this reason, the utilization factor has a direct impact on the status of the queues despite not being one of the six characteristics listed by Kendall and Lee.

$$\rho = \frac{E(X)}{E(t)}$$

An example illustrating this behavior and is presented below in the book develops Quantitative Methods for Managers and Liberman83

Table 2. Using the system.

P	Tail length	Relative F%	F cumulative%	% Causes	cut	Absolute value
0.9	9	42.5	42.5	12.5	55.027	45.0
0.8	4	18.9	61.4	25.0	86.428	13.6
0.75	3	14.2	75.6	37.5	113.104	13.1
0.7	2,33	11.0	86.6	50.0	136.613	36.6
0.6	1.5	7.1	93.7	62.5	156.201	56.2
0.5	1	4.7	98.4	75.0	173.426	73.4
0.25	0,333	1.6	100.0	87.5	187,500	87.5
0	0	0.0	100.0	100.0	200,000	100.0

The table above presents the utilization factor and the queue length of a M/M/1 system equal to rho lambda, a single server and service rate equal to unity. In addition, this table shows the cumulative relative frequency and the total number of people in queue.

This information is used to apply the Pareto principle or rule 80/2084, which states that 20% of an action will produce 80% of the effect, while the remaining 80% will result in only 20% of the effects. In applying this principle it is possible to define at what percentage use the highest concentration of people queued is generated.

The result shows that 37.5% of said ρ represent 75.6% of the queue length. Therefore, the system uses were taken within the range from 0.7 to 0.99, constant jumps by a decimal unit, not taking into account levels equal or greater than 1 use because the model remains in its stable state * to 0.9985; thereafter the system would be in transient condition and would not be possible to perform the analysis we want to research.

5.11.16. Service Time Distribution and Service Fee Per Node

One objective of this study is to determine how the variation of time between arrivals impact system performance, for this reason decide to raise and analyze two scenarios compared with different probability distributions of inter-arrival times. Accordingly, in all scenarios to be generated, different probability distributions for inter-arrival times, either exponential or hyper Erlang be taken; and service times equal rates constant attention will be handled.

For the definition of the probability distribution used to model the service times, it is taken as reference Exponential distribution, as this is the most common within the queue theory possible to model independent processes for their property of non-memory, enabling service time of a customer who arrived at a certain time is distributed exponentially as time another customer arrives after a period, mutually independent.

After defining the probability distribution for service times, we must then determine their average attendance rate, which represents the expected number of people served per unit time in the system.

To set this service fee it shall be taken as reference grade work by Sandra Alvarado Vera entitled "Design of a strategy for improving the quality of service workshops through technical training based analysis General Motors and in this study an automotive workshop is presented with several points of care or queuing systems connected in series with a probability distribution between exponential service time and a variable number of servers per node.

By sampling in such work variable rate care for each point it was calculated. Service fees calculated, the lowest rate is selected, as the said point being part of a series of processes for repairing an automobile, this rate that determines the level of care system. It is important to remember that according to the theory of constraints in serial processes the level of care will be determined.

Due to the above analysis, the μ rate customer service 0,16 per minute calculated Vera Alvarado's work takes as input to this study, not only for compatibility between studies but also as the most important reference mark the pace of attention of a whole system, even though there are other average rates of different proportions.

5.11.17. System Capacity

System capacity is one of the most influential in the perception of customer service to quality factors. If the capacity is insufficient, a number of customers remain unattended even to enter the place of care. On the other hand, those who were on it when it is at full capacity probably will feel uncomfortable and dissatisfied.

Such capacity must therefore be consistent with the demand and supply service and limited by the physical space available. In the present study the capacity of each node will be determined by reference to service centers of different sizes and sectors located in Bogota. Three different levels as follows be taken into account.

5.12. EXPERIMENT DESIGN

Each of the factors that influence the behavior of queuing networks and have been described previously result in an experiment that will provide the information needed to perform the tests generate results and then analyze the impact on performance measures the networks will generate.

Then the design of the experiment is presented following the methodology proposed in the book Statistical Principles for Design and Analysis Research by Robert Kuehl under a specific protocol that gives the possibility to observe the implications of the important results

5.12.1.The Purpose of the Study

Determining the performance of different networks tails generated by combining features of the waiting lines as probability distributions of inter-arrival times, number of nodes, servers node in the system, the utilization rate and system capacity.

In addition, the analysis will seek to find systems generated values to improve the performance of the systems analyzed.

5.12.2.Response Variables

The key measures of performance of a queuing system are the expected number of customers in the system (L), the number of entities in the queue (Lq), the expected time the (W) system and the expected time in queue (Wq). Of these performance measures will be selected as the response variable Lq and Wq, as they are showing how such behavior queues.

Due to the absence of bounded formulas to easily calculate performance measures of queuing networks it is necessary to use advanced calculation techniques. For this simulation will use the Rockwell Arena software, which generates consolidated reports for all runs and detailed reports for each replica.

To know the response measures models, detailed data report generated in Microsoft Access, as shown below will be taken:

The information is obtained from each simulation model will be averaged and consolidated in Microsoft Excel, where the confidence intervals that will do further analysis will be generated.

The confidence interval or Half width is under the assumptions of the central limit theorem, since the number of observations is suffering large that the data have normal behavior.

The formula for calculating is:

$$Half\ Width = Z1 - \alpha / 2$$

5.12.3.Controllable Factors

In short, to see the behavior of queuing networks the following controllable factors vary:

- Number of network nodes;
- Number of servers per node;
- Probability distribution of inter-arrival system;
- Utilization factor per node;
- Capacity of each node.

With these two main systems will be formed: one exponential probability distribution for inter-arrival times, time distribution exponentially attention

FIFO queue discipline, finite capacity and infinite target population (M/M/s/k); and one with Hyper-Erlang distribution inter-arrival times, time distribution exponentially attention FIFO queuing discipline, finite capacity and infinite target population (G/M/s/k).

5.12.4. Type Measurements/Observations

Observations specifically concentrate on measuring indicators Lq and Wq. Because these measures are averages of time and entities in the queuing system are continuous type and greater than zero.

5.12.5. Factor Levels

Table 3. Summary of levels factors.

Factor	levels	Description
Number of network nodes	3	1, 2,3.
Number of servers per node	6	1,2,3,5,7,10
Probability distributions	2	Exponential, Erlang hyper-
inter-arrival times		
Utilization factor	4	0.7, 0.8, 0.9, 0.99
Capacity of each node	3	10, 20, 30

5.12.6. Size Experiment

The nature of networks and the number of factors and interactions of these enable comparisons between treatments are affected substantially by the conditions in which they occur. Because of this, a factorial design is the methodology and appropriate treatment for this experiment, as each observation provides information on all the factors and may view responses of a factor at different levels of other factors in the same experiment 94.

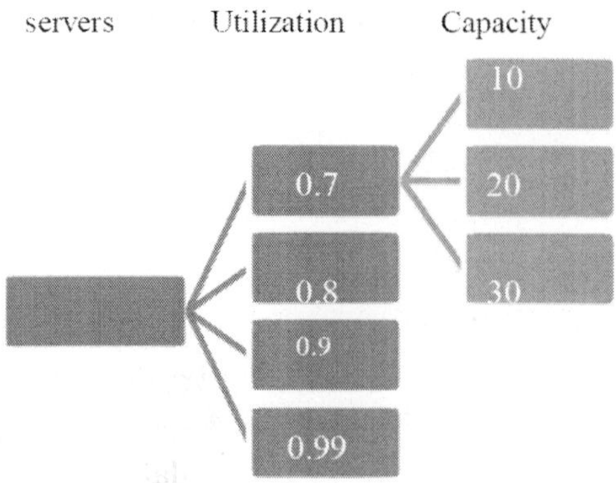

Figure 6. Factors Nested

Besides being a factorial design, have a nested experiment, which means that each particular level of one factor occurs at all levels of nested another factor, as shown in Figure 5.6. Thus, the number of combinations of design is nested multiplication factor by interaction in each of the nodes. (Table 5.4).

Table 4. Number of experiment combinations

Model	**Amount of**		**Amount of**	**Amount of**	**Amount of**	**combina-tion**
	Distribution		**utilizations**	**capabilities**	**servers**	**ones**
1 Node	2		4	3	6	144
2 Node	2	4		3^2	6^2	2592
3 Nodes	2		4	3^3	6^3	46656

A total of 49,392 have combinations. Despite the high number of combinations will proceed to make full factorial, since no homogeneity between factor levels as building blocks that allow to generate a fractional factorial model; also not known beforehand interactions that do not make sense or analyze interest to be removed from the full factorial.

With this is terminated the design of the experiment, after carefully define the factors affecting a queuing model network can run the simulation in order to produce conclusive results of the behavior of the queues and make an analysis to decode an improvement method for such systems.

5.13. SIMULATION EXPERIMENT

Determining the impact factors into queues made above produce oriented simulation due to high difficulty mathematical models approach. This chapter will explain the procedure, size and configuration that it must have to obtain the expected results.

5.13.1.Setting the Simulation

The simulation was performed using the software Rockwell Arena in three different models depending on the number of nodes that we want to simulate. Due to the large number of systems to be simulated, it performed a macro to run the models automatically from different combinations of factors that are organized into a Microsoft Excel file (see Annex A). This macro automates the process by taking the input information from Excel and take it to Arena to run each of the 49,392 possible combinations.

5.13.2.Definition and Duration Replication

The number of replicates is determined with the aim of simulating a year of work in the service sector. Companies within this sector work Monday through Friday eight hours according to the labor law; and weekdays for 2010 according to Article 62 of Law 4th of 1913 (Art. 70 CC) 96 are 252.

5.13.3.Variance Reduction Techniques

To find the experimental design efficiency and improve accuracy, there are techniques that seek to reduce the variance and give better use of information available. Such techniques are used in statistical studies where it is necessary to obtain results with the lowest possible variability to and rightly conclude accurate comparisons.

One such technique applicable to this study is the Monte Carlo method, since it allows performing experiments using pseudo-random numbers. The advantage of the Monte Carlo method for this study is that it has an absolute error of estimate that decreases as that the sample size is increased the experiment, i.e., that the greater the number of replicas will be less variance thereof.

To apply this method in the present study, we resort to methods of generating random variables with the intention of expressing the probability distributions based on uniform random numbers. Using the method of the inverse transformed to the exponential distribution has to:

$$F(x) = 1 - e^{-\gamma x}$$

$$1 - e^{-\gamma x} = R$$

$$xe = -\frac{1}{\gamma}\ln R$$

Where xe are exponential random variables.

In the case of Hyper-Erlang distribution, the inverse transform sampling, despite being the most common and simple procedure can not be used since it is not possible clearance of the random variable. Due to this, it uses the convolution method and the same properties to generate distribution of the random variable values.

By the definition of the distribution it is known that this is the sum of Erlang random variables each with expected value C/λ. Therefore, as in the case of the Erlang distribution is possible to generate random values with the sum of n random exponential variables, Hyper-Erlang can be generated with the sum of random variables Erlang.

It is known by the convolution method that a random Erlang variable is given by:

$$x = -\frac{1}{\gamma}\ln\prod_{i=1}^{n} Ri$$

Then, the sum of these serious variables:

$$x_{he} = \sum_{i=1}^{n} x_i$$

Where xi are random variables Erlang. Accordingly xhe represents the generator of random variables Hiper-Erlang.

Now, in order to increase the accuracy of the experiment in the case of the latter distribution, it is decided to apply another method of reducing variance that is compatible with the already used, called antithetic variables. This procedure is based on introducing negative correlation between two consecutive samples in order to reduce the total variance.

In the case of Hyper-Erlang distribution defined in the experiment has to be constant to each phase lambda, the variance calculated above XHE generator is given by:

$$Var(x_{he}) = \frac{1}{N}(\sum_{i=1}^{N} Var(Ri) + 2\sum\nolimits_{j=1,j<>i}^{n} Cov(Ri, Rj))$$

So that if possible the covariance between the random variables Ri, j is negative, the total variability of replication is reduced.

Since Ri, j are numbers uniformly distributed between zero and one, such that $Rij \sim U\ (0,1)$ then (1-R) also uniformly distributed. As both are unbiased estimators, one is the complement of the other and their covariance is negative are called antithetic variables.

For this case, the easiest way to get negative covariance between the six pseudo-uniform random numbers included in the estimator XHE is that R2i-1 variables take the value of R, and R2i variables take the value (1-R).

Accordingly, it results in the generator of random variables Hyper-Erlang the following expression:

$$x_{he} = -\frac{1}{\gamma}\ (\ln(R(1-R)R) * ((1-R)R(1-R))\ )$$

5.13.4 Time Simulation Execution

In order to estimate the execution time of the simulation, taken as reference the duration of execution of each replica, which is approximately one second. If 252 replicas will be made to complete a year of

5.13.5.System Requirements

Run each simulation model generates a file in Microsoft Access in which consolidates information. This file weighs on average 5.23 MB, for a total of 258,160 MB for all models. Additionally, each model stores a text file of time between each node outputs a total of 144,648-to files-each of which has a weight average of 410 KB (total: 58785680 KB).

In general, the entire simulation process will generate information with an approximate weight of 310 GB. Accordingly computers with a processor speed equal to or higher than 1.2 GHz, greater memory 1 GB hard disk and minimum capacity of 310 GB is needed. As for the software, the operating system must be Windows and programs are needed as Rockwell Arena, Microsoft Access, Microsoft Excel and Visual Basic with ADO (ActiveX Data Objects) and DAO object (Data Access Object) for the execution of the macro that unifies all information (see Annex B).

5.13.6.Construction Simulation Models

For simulation of queuing networks models representing the experiment and that in turn yield the desired results will be constructed. Such models are produced entirely by the following modules:

- Create for the arrival;
- Dispose for the outputs;
- Process for each node or point of care; and
- Decides to limit the ability of nodes.

Additionally there will be a ReadWrite module for keeping track of inputs and outputs to each of the network nodes queue. Although the study of this behavior is not considered within the scope of the current proposal it can be retained for future study.

5.13.7.Module Configuration

Create: The queuing network will consist of a single input where users arrive according to a time between arrivals Exponential or Hyper-Erlang is distributed. In the case of the latter, as Arena not have this distribution in their libraries, it is necessary to generate random variables follow such distribution from pseudo random numbers. To this end the resulting expression techniques variance reduction to the arrival times as a Hyper-Erlang distribution is used. In the case of the exponential distribution, the exponential distribution even if there is in bookstores simulation software, also calculated in number of variance reduction techniques will be used generator.

Decide: After the login process is a decision module where system capacity is defined, the queue status is monitored and the flow of entities is addressed in the model, either towards the beginning of the process or to its removal system. By NQ (Service Queue) <Node Capacity command user input to the process so that upon reaching the rated capacity not input new users allowed is limited. If users cannot access it, this module routed to an exit system represented by the Dispose module. (Figure 5.7).

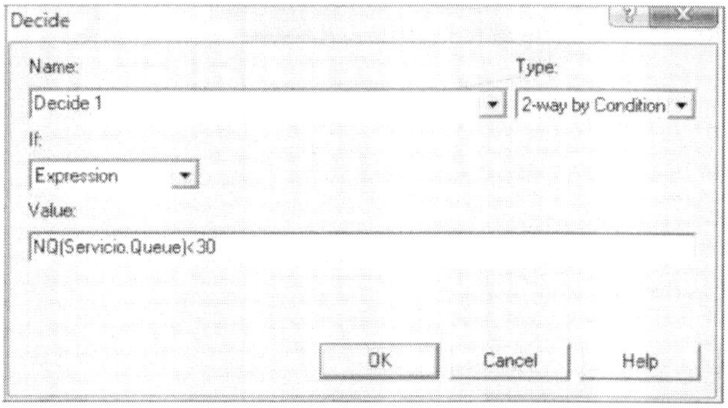

Process: It represents each of the service nodes in the system. This is where entities processing is performed and where there are a number of parallel servers that are assigned to customers according to the rule Smallest Number Busy. The service time follows exponential distribution with parameter μ remains constant in each of the nodes.

ReadWrite: Finally there is a module ReadWrite which it will record the current time of each process output and the start of the next transaction service by Tnow attribute. The purpose of this module is to keep track of the number of outputs and times between them in each of the nodes along the simulation.

Dispose: It is the module that represents the output of the entities queuing system. The outputs can occur in two different moments: the first is when the capacity of the nodes or points of care is inadequate for the number of people who want to access it. At that time the system is full and the module decides entities routed to some Dispose. The second output is when an entity managed to enter the system reaches the end of the queuing network and leaves the system.

5.13.8 Simulation models

Each simulation model is composed of arrival, the nodes along the length of pattern (1,2 and 3) and finally output.

5.14. EXPERIMENTAL RESULTS

After performing the simulation and calculation of the performance indicators of the models posed, it is necessary to analyze systems and determine how the variation of the characteristics of the queuing network considered here can achieve better results. With this analysis, we can glimpse the main differences between the two selected to determine the change and the impacts between the scenarios studied distributions. For this analysis we proceed to apply statistical tests using the SPSS program on the 49,393 consolidated information models of the experiment (see consolidated database in Appendix C). Through this analysis we can check evidenced relations through a descriptive analysis.

In the tests performed this homogeneity of variance or ANOVA measuring the influence of categorical variables (factors) on the dependent variable quantitative, in this case on Lq and Wq; and evaluates the main effects of these factors and interrelationships. In addition to this test, a study analysis of homogeneous groups and confidence intervals interactions between factors complements.

Then the simulation analysis, which seeks to determine by statistical procedures how a queuing model behaves network according to the factors discussed in the experimental design and extract their results key information seeking to improve the overall efficiency is presented systems developed here.

5.14.1.General: Interactions between Factors

With the results we can see the difference between marginal means of the factors and within view individual relationships (first grade) and combined (second and third degree) factors on the performance indices. Less than 5% significance level of p value interrelation of factors confirm the results.

Within this interaction relations second order Lq and Wq are:

- Utilization factor and number of servers (only for Wq);
- Factor utilization and system capacity;
- Number of servers and system capacity;
- Distribution of inter-arrival times and utilization factor;
- Distribution of inter-arrival times and system capacity; and
- Distribution of inter-arrival times and number of servers.

Third-order interactions for Lq and Wq are:

- Utilization factor, distribution and capacity; and
- Utilization factor, capacity and servers.

Interactions between factors as levels of utilization, and distribution servers; or level of utilization, distribution servers and capacity have not significantly associated with the results, this means that although individually these factors impact the results of the dependent variables (Lq and Wq), when the level of interaction increases with combination of factors but its effect becomes less important.

5.14.2. Analysis of Average Number of People in Systems

The expected number of people queued is one of the two selected performance indicators to measure the performance of queuing networks. Within the design of the experiment all the variables that will be considered to affect the fluctuation of this indicator are marked. Then, through the organization and study information, a descriptive analysis of behavior will be made evident.

5. 14.3. Relationship between the Expected Number of People Queued and Node Attention

The expected number of people under input parameters and endless attention and capabilities such should be the same in each of the nodes as Jackson and Burke's theorem and in systems with exponential arrival times and services. However, in systems with finite capacities average output rate from one node to another differs from the entrance fee. This behavior occurs because not all entities who wanted to access the service could do so, and therefore abandoned the system by lowering the rate to the next node.

This situation causes the arrival rate to the following nodes in the network is lower and the expected number of people queued decrease. (Figure 5.8).

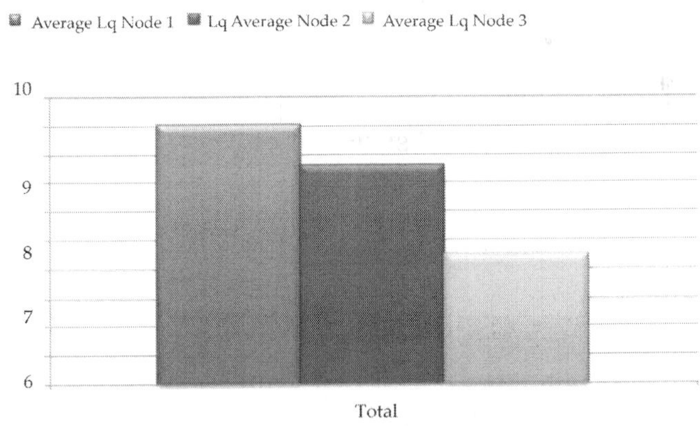

Figure 5.8. Results of Lq according to the node attention Exponential time between arrivals

As in systems with exponential times between arrivals, in the case of Hyper-Erlang distribution, the arrival rate changes and make the first node tends to congest. To the extent that the node is farthest from the input source, Lq values will be lower. Figure 18 evidence this behavior between node and network node. (Figure 5.9).

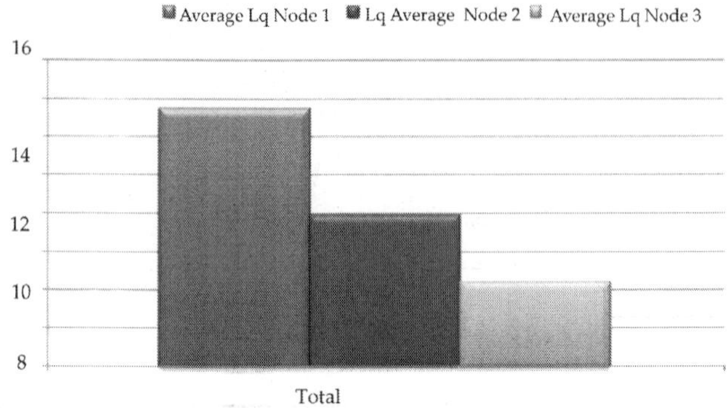

Figure 5.9. Results Lq as the node under Hyper-Erlang distribution

The difference between the first node and subsequent nodes is that the arrival rate and the distribution of the inter-arrival times make the first focus long lines to prevent the passage to the following nodes are formed, formed a neck bottle. The rate of output of the first node to the next is substantially reduced and this effect continues chain becoming even more evident as the saturation increases in the first destination station and the network is longer.

Making a comparison of the results of Lq for both distributions are greater changes to 44% of the Hyper-Erlang regarding exponential, whose variation is more representative in each node due to the greater variability in the results of the Hyper-Erlang at the entry point of the front of the exponential system. (Table 5.5)

Table 5. Switching between distributions along the length of the network (lq)

Length	Lq
1 Node	44.4%
2 Node	7.83%
3 Nodes	5.40%
Avg. Total	19.21%

5.14.4.Relationship Between the Expected Number of People Queued and Utilization Factor

The utilization factor quantitatively describes how the system is saturated. It is therefore natural to see that their increase also impacts the number of people in queue (Lq). As seen in Figure 5.10, the performance indicator Lq increases as the utilization factor is about one only for the first node is where the value is known.

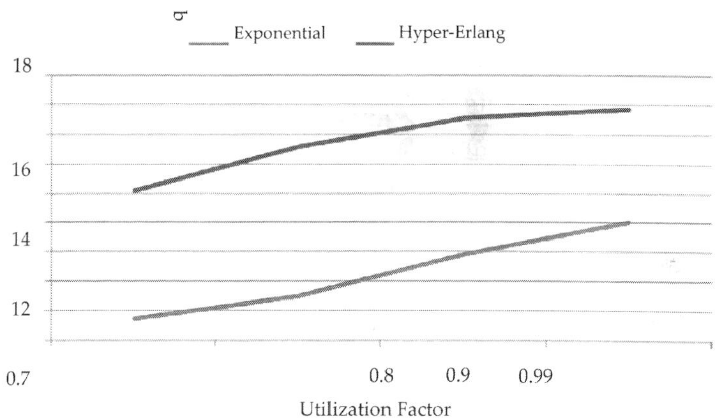

Figure 5.10. Results of Lq according to the utilization factor in the exponential distribution in the first node.

For systems with exponential distributions, growth Lq relative to Rho is linear, with a slight change in slope makes the indicator value is increased when approaching the utilization factor one. For Hyper-Erlang distribution, the Lq is much greater with a concave curve also increases as the Rho it approaches one.

Changes between distributions are much more tangible when the utilization factor is less and exhibit significant decreases as the Rho approaches one. Saturating the system makes the Hyper-Erlang lose acceleration in growth Lq while still exponential increase linearly. (Table 5.6).

Table 5.6. Switching Between Distributions According to The Utilization Factor (Lq)

Utilization Factor	Lq
0.7	426.35%
0.8	401.43%
0.9	146.22%
0.99	25.75%
Avg. Total	249.37%

5.14.5.Relationship Between the Expected Number of People in Queue and Number of Servers

As the number of servers increases in the system, the ability to leave the rows becomes tangible, especially when the variation server moves from one to two. This means that the impact of the variation of a server in a row is higher among fewer servers have at that time in the node attention.

Figure 5.11 shows the changes between the two distributions compared. Evidently the values the indicator Lq Hyper-Erlang distribution are much higher. For practical purposes, the significance of this variation indicates that so wrong could be a model if a distribution that does not really represent what happens in a spotlight could be chosen; for example, in systems with a single server selection distribution inter-arrival times it generates a change Lq of 33.4%; and even more drastic change is when there are 10 servers in the system, where Lq 84.11% would increase. (Table 5.7)

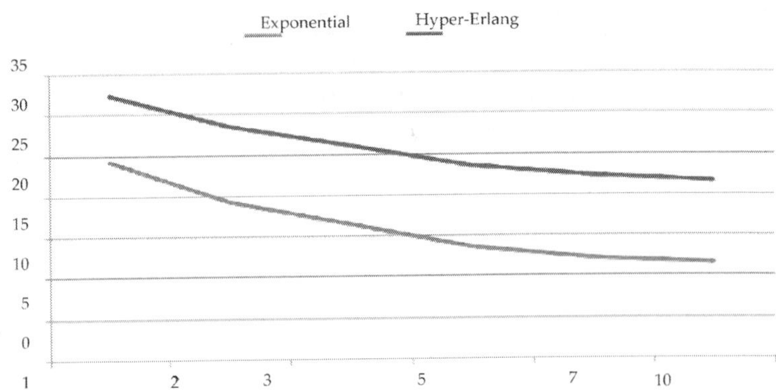

Figure 5.11. Comparative distributions according to the number of servers

Table 5.7. Change Between Distributions According to the Number of Servers (Lq)

Servers		Lq
1		33.40%
2		47.10%
3		56.88%
5		71.32%
7		80.28%
10		84.11%
Avg. Total		57.34%

5.14.6.Relationship Between the Expected Number of People Queued and Node Capacity

Node capacity limits the number of people who trust in him, therefore, the greater capacity, the greater the average number of people in the system. However, when utilization is 0.7, the impact of the capacity is minimal and in some models is indifferent to the number of people queued, meaning that no matter what capacity they have at the point of care the number of people who expect there will be similar across all systems. If this happens with Rho 0.7, it is possible to conclude that use lower levels of this behavior should be similar and even more evident, therefore, not be important to study factors less use 0.7 as the serious Lq indifferent in any capacity.

The changes between the two distributions tend to remain constant as shown in Figure 21, about 57.34% on average. The most significant difference is with the capacity of 10 entities where the value of Lq for Hyper-Erlang distribution changes with respect to exponential in 67.48%, and the capacity of 20 to 30 with a variation of 57.80% and 48.97%, respectively. (Figure 5.12) (Table 5.8)

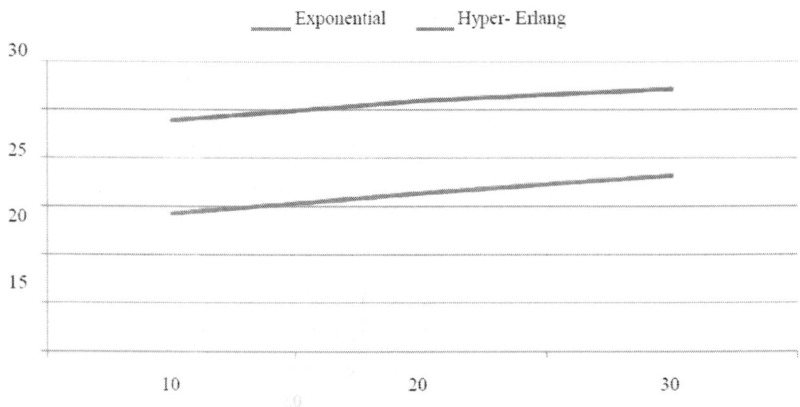

Figure 5.12. Comparative distributions for the capacity (Lq).

Table 5.8. Comparative Distributions for The Capacity (Lq)

Capacity	Lq
10	67.48%
Twenty	57.80%
30	48.97%
Avg. Total	
	57.34%

5.14.7.Relationship Between the Expected Number of People in The Queue, Number of Servers and Utilization Level

Generally speaking, the variation of Lq relative to the number of servers and the level of use no direct interaction with values Lq, however their relationship within the first node if allows to demonstrate the effects of the factors in the onset of network queue.

For the case of exponential distribution models, the Lq grows apace as rho increases when up to 3 servers, and linearly thereafter for each of the nodes, as shown in Figure 5.13.

The results with Hyper-Erlang ratify the lower impact of increased rows servers when debugging as there is more of the same at the point of care.

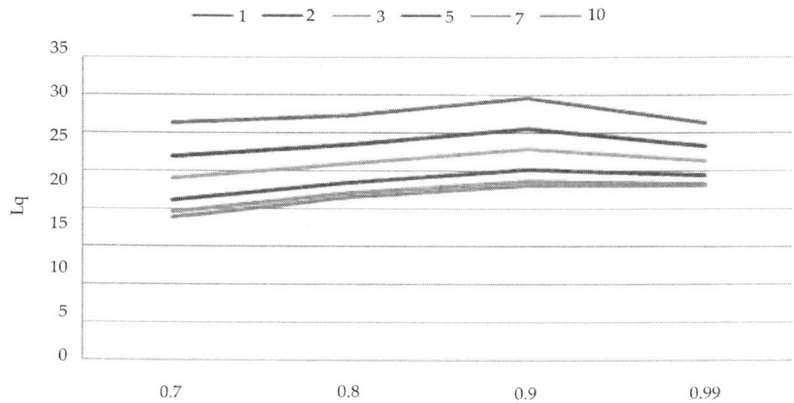

Figure 5.13. Results of Lq on the first node based on the number of servers under Hyper-Erlang distribution

Additionally, the results with hyper-Erlang distribution presents a turning point to match utilization levels of 0.9. This event is due to the saturation of the system to reach the limit of 30 people per node. At that time although the trend say that Lq should continue to rise, it stagnates then left not enter or continue in the system to all entities generated by the model. Thereafter, Lq decreases dislodging rows and serving people who remain in the system.

5.14.8.Relationship Between the Expected Number of People in the Queue, and Node Capacity Utilization Level

Lq variation regarding the capacity utilization level and show whether or not to possess greater capacity to accommodate customers according to workload has a network of service points.

In models with exponential distribution, capacities 20 and 30 people per node is almost indifferent to the result of Lq on the use of 0.7 and 0.8. To the extent that the use grow curves representing Lq in different capacities tend to become convex the greater system capacity.

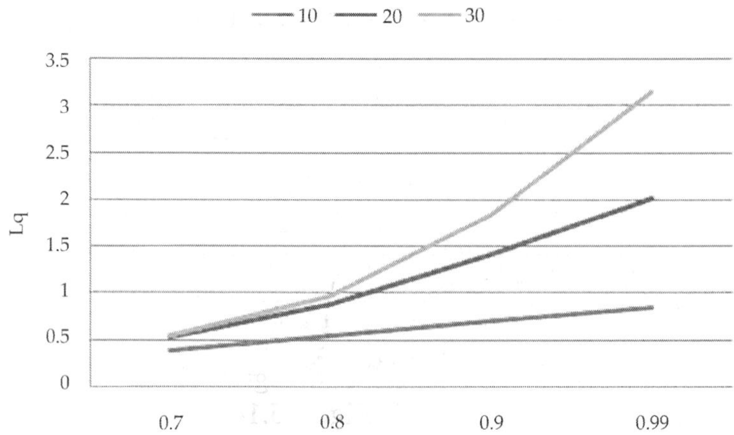

Figure 5.14. Results of Lq according to the first node capacity under exponential.

Although the effect described above is similar for the two probability distributions, a difference related to the growth rate, since the behavior of the curves under Hyper-Erlang distribution tend to be concave (Figure 5.14). This phenomenon does not apply when the capacity is 10 people, where Lq grows linearly regardless of the distribution.

There is evidence of unusual situations, showing erratic behavior from or above 0.9 Rho. The explanation for this is the same phenomenon as explained in Figure 5.15, wherein collapses attention, and mass retirements occur. This creates the turning point when the capacity is equal to 30 entities.

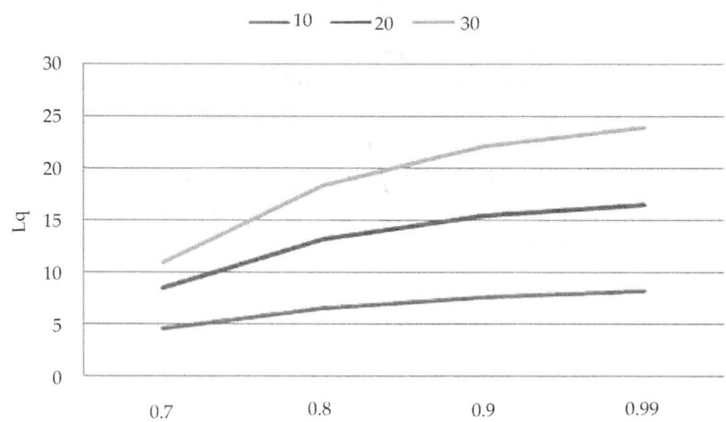

Figure 5.15. Results Lq according to the capacity on the first node under the Hyper-Erlang distribution.

5.14.9.Relationship Between the Expected Number of Queued Node Capacity and Quantity of Servers

Generally speaking, the variation of Lq on the amount of servers decreases as the latter increases for each of the capabilities. Curves representing capacities tend to approach as there are more servers until they join about 7 servers; thereafter indifferent variation Lq, which demonstrated that increased over ten servers, in this case would have no effect on the queues.

For systems with exponential distribution, it can be observed to the extent that the servers are approaching seven is indifferent system capacity; whereas when there is only one server, the greater system capacity will increase dramatically Lq as shown in Figure 5.16.

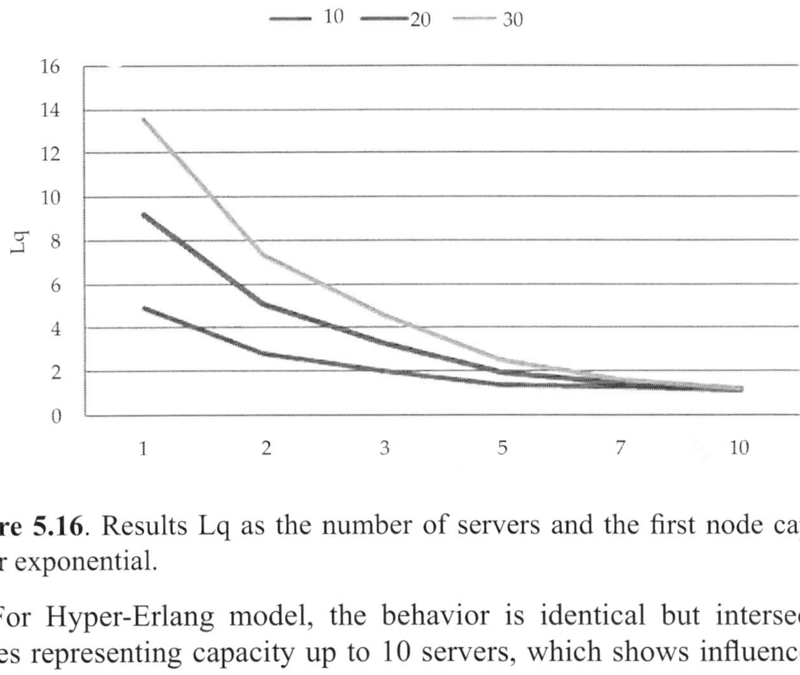

Figure 5.16. Results Lq as the number of servers and the first node capacity under exponential.

For Hyper-Erlang model, the behavior is identical but intersect the curves representing capacity up to 10 servers, which shows influences the ability of the value of Lq for all levels of servers except the latest.

Regarding Lq values the percent change in the results of the Hyper-Erlang regarding exponential averages of 57.34% for each of the servers according to their capacity. (Figure 5.17).

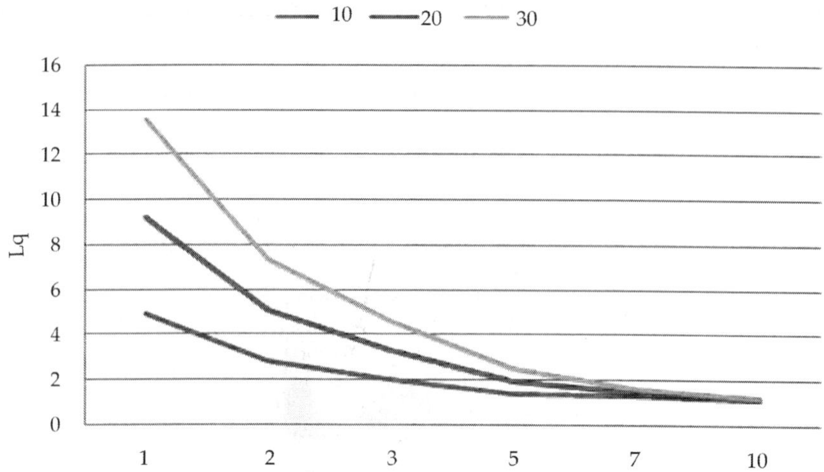

Figure 5.17. Results Lq as the number of servers and capacity on the first node under the Erlang distribution hyper.

5.14.10. Summary of Interaction of Factors on The Expected Number of People Queued

Generally speaking, the variation of Lq on the amount of servers decreases as the latter increases for each of the capabilities. Curves representing capacities tend to approach as there are more servers until they join about 7 servers; thereafter indifferent variation Lq, which demonstrated that increased over ten servers, in this case would have no effect on the queues.

For systems with exponential distribution, it can be observed to the extent that the servers are approaching seven is indifferent system capacity; whereas when there is only one server, the greater system capacity will increase dramatically Lq as shown in Figure 17.

Table 5.9. Summary Relationship between Expected Number of People Queued and Node Attention.

Relationship distribution	Exponential distribution	Hyper-Erlang
nodes	nodes: Lq decline slowing as the node is farthest from the input source. Variation node one to two less than the twothe third node.	nodes: Lq decline slowing as the node is farthest from the input source. greater variation between nodes and especiallyfirst to the second node

Table 5.10. Summary Relationship Expected Number of People Queued and Utilization Factor

Relationship	Exponential distribution	Hyper-Erlang distribution
F. Utilization Lq	Node 1: Increase linear of the Lq with trend to accentuated when the Rho approaches 1. expected average maximum number of people in row Out of 8.	Node 1: Lq increase slowing as the Rho approaches 1. Minimum people in row

Table 5.11. Summary Relationship Expected Number of People in Line and Node Capacity

Relationship	Exponential distribution	Hyper-Erlang distribution
Capacity	Node 1: Lq capacity increases as does, except using 0.7. Growth becomes convex capacities 20 and 30; and fully linear capacity 10people. Node 2 and 3: Lq with little pronounced linear increases and tend to remain constant. .	Node 1: Lq increases as capacity so does a logarithmic curve. Node 2 and 3: Lq linear behavior in all capacities with dramatic inflections when the system reaches its limit care.

TIME AVERAGE OF WAITING IN QUEUE

Perceived most important indicator for the customer is the waiting time before being atendido104. This indicator is linked to the quality of service and is a reflection of how well structured is the process to the public. Wq behavior is similar to Lq indicator described above. Then new interactions between factors and conclusions about the performance of queuing networks are displayed.

5.14.12. Relationship Between Expected Time and Queue Node Attention

The average wait time according to the node attention presents the behavior mime with Lq indicator, the graph gives evidence that while changes have the same tendency differ in value. (Figure 5.18; Table 5.12).

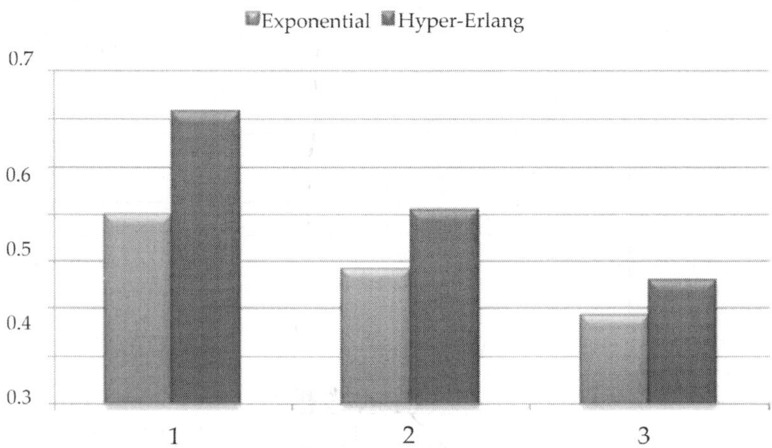

Figure 5.18. Comparative Wq change along the nodes between the distributions.

Table 5.12. Comparative Distributions According to the Length (Wq).

Length	WQ
1 Node	54.32%
2 Node	43.98%
3 Nodes	40.56%
Avg. Total	45.62

5.14.13. Relationship Between Expected Time in Queue and the Utilization Factor

The average wait time in queue is directly related to the utilization factor because as long as the system is fuller longer people wait to be served. Figure 5.19 shows the behavior of Wq by type of distribution, making it clear that the use of one of the two distributions determine different system states. In the case of the exponential distribution, the waiting time at the first node increases linearly. Meanwhile, in the Hyper-Erlang is a pronounced variation Wq from the level of system utilization. It is perceived saturation Rho equal to 0.8 where there exits the system (Figure 5.19).

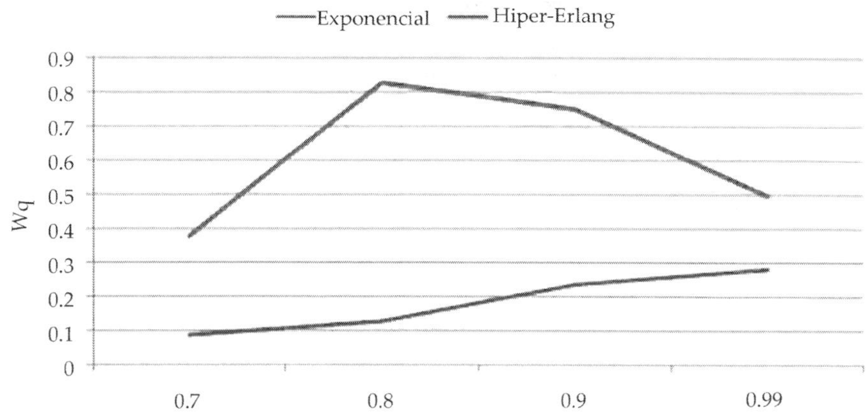

Figure 5.19. Results of Wq on the first node as the utilization factor.

Hyper-distribution Erlang is observed that when it comes to saturation, the system appears to behave atypically due to congestion presenting point

Attention. The system outputs are massive, which makes Wq decrease drastically.

The percentage change of Hyper-Erlang regarding exponential reaches a maximum level of use of 0.8, with a range of 575.91% and a minimum at 0.99 with a change of 69.08%. (Table 5.13)

Table 5.13. Comparative Distributions According to the Utilization Factor (Wq)

Utilization Factor	WQ
0.7	287.93%
0.8	575.91%
0.9	204.09%
0.99	69.08%
Avg. Total	284.47%

5.14.14. Relationship Between Expected Time in Queue and Number of Servers

The number of servers in the service points determines the average waiting time in the queue, the greater the amount of time servers decreases along of all the factors of use. This behavior becomes even more evident when the system has a use close to one. Furthermore, the number of servers has a positive effect on the clearance of queues especially when there are fewer servers in the system.

As was evident in the relationship of Lq, both distributions differ in their results and this difference increases as the number of servers increases; However, the difference with just one server distribution with respect to each other is much smaller than previously seen. (Table 5.14).

Table 5. 14. Comparative Distributions According to the Number of Servers (Wq)

Utilization Factor	WQ
0.7	287.93%
0.8	575.91%
0.9	204.09%
0.99	69.08%
Avg. Total	284.47%

5.14.15. Relationship Between Expected Time in Queue and Node Capacity

The waiting time increases as there is more capacity in the nodes, indicating that capacity is determining when to serve a greater number of people. Although the values of Lq and Wq increase as the capacity grows, in some cases insufficient capacity makes the number of people who try to access the system does not increase. To Wq it is evident that the differences between the two distributions are much lower (about half that seen in the analysis of Lq). (Figure 5.20; Table 5.15).

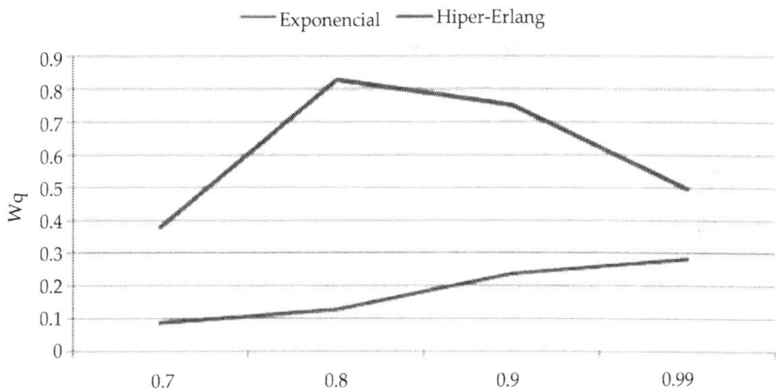

Figure 5.20. Comparative distributions according capacity (Wq).

Table 5.15. Comparative Distributions According Capacity.

Utilization Factor	WQ
0.7	287.93%
0.8	575.91%
0.9	204.09%
0.99	69.08%
Avg. Total	284.47%

5.14.16. Relation between Expected Time in Queue, Number of Servers and Utilization Level

The queuing time decreases significantly as the number of servers is increased, however this impact tends to be more or less noticeable depending on system usage.

In the exponential distribution, the waiting time is lower for Rhos equal to 0.7 and increases rapidly as the use approaches one. A Rho 0.7, there are no variations in Wq As more servers. If factors of less than 0.7 use is likely to take the curves come much more until the amount is indifferent with respect to server Wq.

Additionally, this distribution behavior evidenced in the study of Lq is confirmed: the impact on the waiting time is increased when a server is greater if there are fewer servers at the point of care. (Figure 5.21).

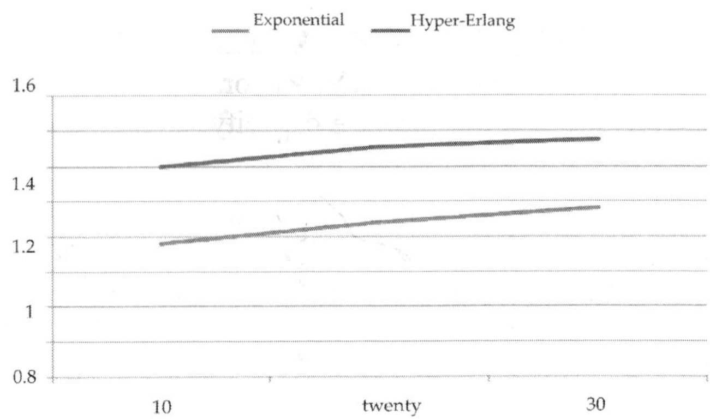

Figure 5.21. Results of Wq in all nodes by the number of servers under the exponential.

In the case of Hyper-Erlang, the impact of change between servers is much lower for one and two servers, and higher for the rest of them with respect to the exponential; Further along the Rho difference between Wq varies less and the trend curves as the number of servers is concave. (Figure 5.22).

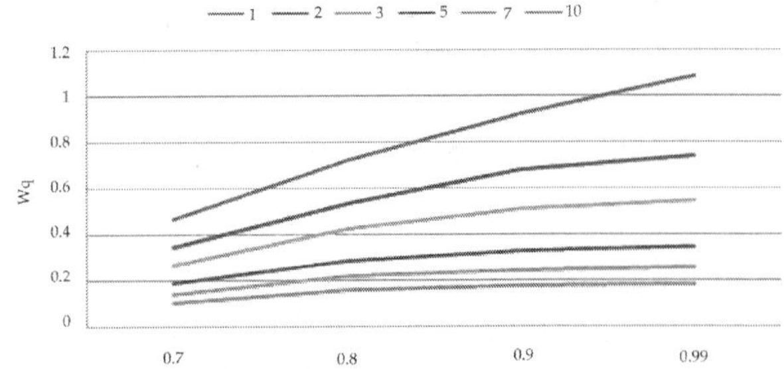

Figure 5.22. Results of Wq on the first node based on the number of servers under Hyper-Erlang distribution.

In the two node change Wq is indifferent when the number of servers approaches ten, showing that there is no relationship between the number of servers and system utilization, i.e., the results of the waiting time for number of older servers to 10 would be the same no matter how many more servers are included.

5.14.17. Relationship between Expected Time in Queue, and Node Capacity Utilization Level

These three variables interact in order to illustrate the effect of the capacity over time as expected den tail utilization factor. Its effect is similar for both distributions as there are points where capacity changes have no effect on the average performance.

Inside the exponential distribution, indifferent regarding Wq capacity with less use factors to 0.7. For systems with utilization levels of 0.8 and capacities 20 and 30 people, we can show similar behaviors Wq. From levels above 0.8, vs. increasing Wq capacity along the utilization factor increases linearly. (Figure 5.23).

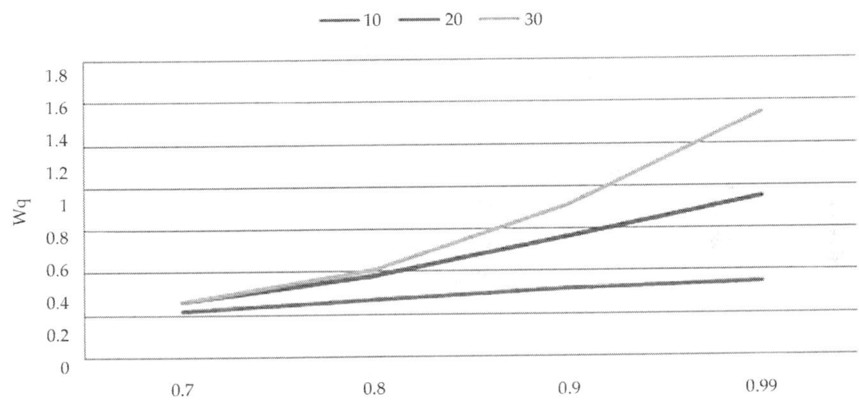

Figure 5.23. Results of Wq by capacity on the first node under the exponential.

Under the Hyper-Erlang distribution, Wq has a concave behavior, which increasingly emphasizes the extent that the utilization factor approaches one. With less than 0.7 Rhos, lines tend to join, showing that no matter what capacity, the waiting time is constant. Atypical movements are due to the arrival of the capacity limits. (Figure 5.24).

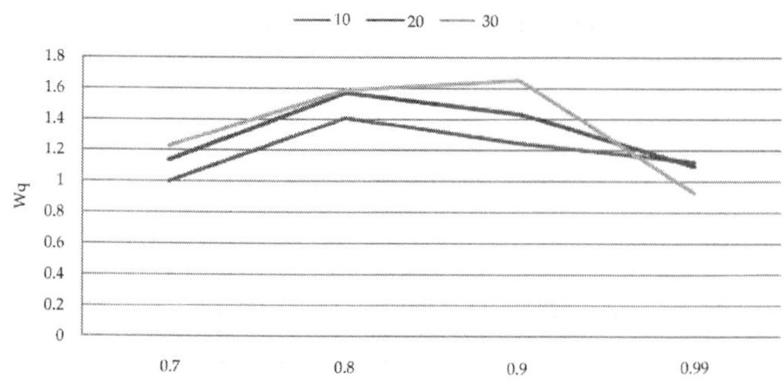

Figure 5.24. Results of Wq by capacity on the first node under the Erlang distribution hyper.

5.14.18. Relationship between Expected Time in Queue Node Capacity and Number of Servers

Increasing the number of servers waiting time decreases while the increased capacity makes expected to increase due to the greater number of people with the possibility of row. However, the difference in values relative to Wq capacity decreases and tend to be the same when the system has 10 servers at the point of care. This effect occurs because at the time the system is relatively clear capacity remains a limitation, that is, that the expected time in the queue is indifferent to the variation in system capacity.

In systems with exponential times, varying between 7 and 10 servers for all capacities it is indifferent to Wq. When the number of servers is lower, the capacity has an effect on waiting time tends to be more reliable especially when there is a single server. (Figure 5.25).

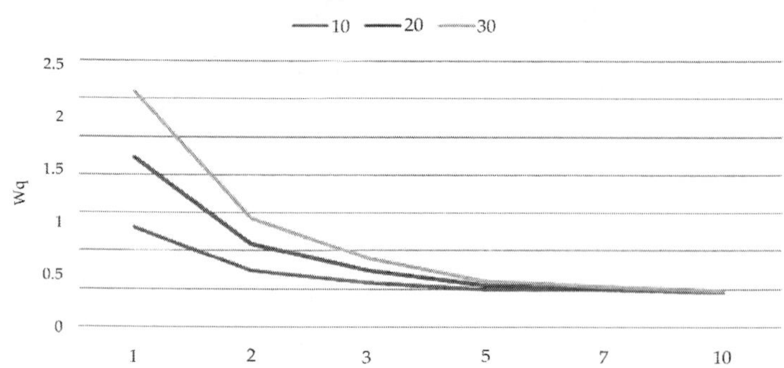

Figure 5.25. Wq results according to the number of servers and capacity in all nodes under the exp. Distribution.

In Hyper-Erlang distribution, it is observed that the node one Wq values decrease as the number of servers. Capacity 20 and 30 closely approximate over the servers, however only up to 10.07 capacity servers without importance in the outcome of Wq.

Wq results in nodes two and three, tend to behave as a convex curve as servers increases. Wq difference with one and two servers is not as great as it was in the case of exponential, if it happens that the capacity does not affect the performance of the queue when there are more than seven servers. (Figure 5.26).

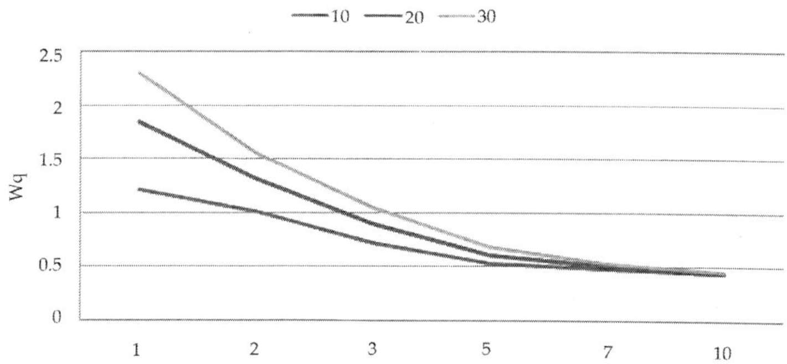

Figure 5.26. Results of Wq by the number of servers and capacity on the first node under the Erlang distribution hyper.

5.14.19. Summary of Interaction of Factors over the Expected Time in Queue

Table 5.16. Summary Relationship Queued Expected Time and Attention Node

servers	WQ
1	18.62%
2	47.63%
3	56.61%
5	71.83%
7	75.50%
10	75.59%
Avg. Total	48.40%

Table 5.17. Summary Relationship Queued Expected Time and The Utilization Factor

Capacity	WQ
10	58.43%
twenty	48.50%
30	40.43%
Avg. Total	48.40%

5.15. GENERAL ANALYSIS SYSTEM

So far we have analyzed the node-to-node queues, however is even more sense to analyze the behavior of the response variables, Lq and Wq in the entire system. This analysis allows to study the effect of nested factors, since it is possible to see configurations of a node according to a different level of the following nodes; further shows the integral relationship between factors within the synergy in a network and check its effect and impact individually on nodes and generally across the network.

5.15.1.Homogenous Subsets

To identify homogeneous subsets Scheffe test for each factor applies. It is found that there are no significant differences between means the null hypothesis with a significance of 5% is rejected.

Test for multiple comparisons of group means. SPSS embodiment.

Depending on the capacity:

When the general capabilities of the system are higher than 30 people, that is, when the capacity at each node is greater than 10 entities sets are identified without difference in their averages capable systems (40, 50, 70, 90) entities.

Depending on the number of servers:

In the case of the number of servers, there are 11 homogeneous subsets, which indicates that no server configuration is independent and they are all related in some way.

The appearance of multiple homogeneous subsets between capacity levels and test servers within a system there are one or more configurations with the same result in Lq and Wq but economic indicators, procedural or different physical impact.

5.15.2. Summary of Performance Measures

The analysis thus far has been developed based on the results after the simulation showed the combination of factors and levels defined for this study. The probability distributions defined for inter-arrival times allowed for comparisons of the performance measures of system interactions, performance measures for Hiper-Erlang times being always higher. In addition, it was possible to show similarities regarding changes, patterns and trends in both scenarios analyzed, which suggests that the findings of this experiment can be extended to other probability distributions, since the behavior for both distributions was homogeneous with respect to effect of factors. Despite this,

5.15.3. Overview

- The higher capacity, greater Lq and Wq.
- The higher level of utilization, the higher the value of Lq and Wq.
- If the number of servers increases, the values of Lq and Wq decrease.
- If Hyper-Erlang distribution is chosen the results of Lq and Wq are much higher compared to the exponential.
- Wq and Lq changes throughout the network nodes in series. This variation is accentuated in the first node and affection lost as it advances on the network.
- Excepting the utilization level, the variation of a factor from first to second level thereof generates greater change in the results of Lq and Wq that variation of these sub levels.

5.15.4. Exponential Distribution

- The values of the indicators tend to have less variability in the results.
- Major changes of Lq and Wq are given in levels above 0.9 used
- System capacity has an effect on Wq and Lq to less than 7 servers.

- The number of servers affects system performance when it passes from one to two servers, in other variations only has effect on the indicators when rho approaches one.

5.15.5.Hyper-Erlang Distribution

- Wq and Lq changes throughout the network nodes in series. This variation is 76% of the first to the second node and the second 12% of the third node. Then, the increased attention on the first nodes will have more impact on the overall system Lq.
- Changes Lq and Wq presented in utilization levels of 0.9 and 0.8 have sometimes extreme situations where the system is not stable.
- System capacity has an effect on Wq and Lq when the utilization factor is close to one or the number of servers is greater than 7.
- The number of servers affects system performance, especially when going from one to two servers.

5.16. DESIGN OF THE OPTIMIZATION MODEL

The results of the simulation models waiting lines outlined different characteristics and conclusions. However, the study has so far only presented information materials and no action plans or developments to improve the already calculated waiting lines statistical indicators. The real impact of this information must be translated into the design and improvement of existing networks of queues in order to mitigate the negative impact they can have on customer perception.

The waiting time of a customer in a row before being seen is considered the indicator most significant quality perception of clients because it shows how well designed the process and what is the actual capacity a company has to meet its necessities. In a globalized world, companies must take into account the quality of their service in order to attract and keep new customers and not assume the costs of losing we already own. Therefore provide the best care is today the number one quality policy in any company.

However, within this new orientation there are objections economic and productive type, for providing optimum quality state where the customer should not wait to be served is an expensive and inefficient option. The resources needed to support this scenario involves too many idle time, therefore, companies de facto rule that option and choose to ignore or postpone the study of this issue. Figure 39 shows the paradox with managers in charge of managing waiting lines: an increase in attention span means extra costs for the company (cost of service); However, if there is a sufficient level of attention a detriment to the quality of service (waiting time) and an economic loss of opportunity and image would be generated, called expected costs. Faced with this paradox, it should seek a balance that meets the needs of customers and meets the possibilities of the company.

5.16.1.Costs Associated with Economically Viable Solutions

The main objective of an optimization model oriented queues is to minimize waiting time and the number of entities in queues. However, minimization of these indicators involves large capital investments by the company, which cannot be ruled out and should be included in the optimization process. Both recruitment in order to debug queues and improve service levels, the adequacy of facilities represent costs that must also be minimized to provide businesses with efficient and economically viable solutions.

Within the definition of the parameters affecting a queuing system, there must be clarity about what it costs implicit, independent nature and cause of the business application are to be taken. These costs will be worked on a monthly basis, and that accounting records are handled with this periodicity and this facilitates the determination of its value.

5.16.2.Labor Costs

Labor is undoubtedly the most valuable in providing a service and also the most expensive resource. Globally, 63% of private sector companies are paying more in personnel costs. The benefits burden of having many employees can lead to bankruptcy any business because within cost unit to produce a good, about 80% represent the raw material.

In addition to salary compensation, costs associated with benefits burden, fiscal contributions and contributions to social security make a worker represents a high cost to the company, in fact when in five Latin American countries the cost of labor up 10%, its share in the export market fell by 2.5 points,

For this reason, labor must be taken into account when designing a system service because their number oversizing generates idle time and cost overruns; and fall below the necessary increase the waiting time, workloads and system efficiency.

Taking into account this, one of the main parameters to improve should aim at efficient allocation of human resources. In network queues defined set, a hand is considered homogeneous and without distinction on the pay employee compensation or servers work.

The total cost of this is equal to:

$$C = CsS$$

where, Cs is the wage compensation servers S and the total of them in the system. It is important to clarify that wage compensation includes base salary, commissions if applicable, social benefits, social security, fiscal contributions, affiliation to occupational hazards, among others.

5.16.3.System Utilization Costs

One of the most important in defining how to provide a service costs are physical facilities. Ray Croc, McDonalds promoter says that the location of an establishment is not only important but it is the most influential factor in the success of a deal. Along with the place many more costs that are derived therefrom and which are necessary to maintain the point of care in appropriate and suitable conditions for the activity associated planned it.

Due to non-characterization of an office or specific point of attention in this study, a total monthly cost, which can be consulted on receipts, invoices etc., defined as follows shall be taken:

$$Ctotal = Car + Cadm + Cman + Crh + Cas + Cim + Csp$$

where Car is the cost of leasing the site, Cadm um represents the cost of administration, Cman maintenance costs, the cost of security Crh, Cas, Cim Csp taxes and utilities. This cost is associated with the number of possible people within the system (servers (S) and capacity customers (K) and is sensitive to variation of them.

Then, to see the costs of changing the capacity or the number of servers that cost per person expressed as follows:

$$C_u = \frac{Ctotal}{s + k}$$

5.16.4.Expected Costs

Waiting for customers in the system is associated with the idle time they lost while waiting to be served. Evaluate this cost is a key factor when quantifying the effect it has on the efficiency of a business. There are companies like Jet Blue Airline quality policy that pays customers to wait a
bonus; However, there are other companies where it is very difficult to calculate that value or cost because of its intangible nature.

The characterization of these costs will depend on each company and is defined by the linear function exposed by Hamdy Taha, where the expected cost is proportional to the size of waiting.

$$Cw = CeLq$$

where Ce is the cost of waiting for each customer.

Therefore, to determine expected costs can use different evaluation criteria, depending on the context, would make tangible costs expected. In other companies it could be calculated based on the opportunity cost involved stop selling an asset when the customer leaves the system or may also be determined by calculating the decrease in demand for products due to deterioration of corporate image It involves having large rows.

5.16.5.Cost of Jobs

Each server that is hired for the proper performance of their duties should have an appropriate job, whose attachment has a monthly maintenance cost, expressed in physical resources, preparation and job induction and office items, among others. Linked to this cost technology today costs are high due monopolies in the generation of software. are classified

$$Cwt = CptS$$

Cpt which represents the monthly cost of having a server in a job. This cost includes monthly value of equipment, monthly value of licenses, stationery value, and monthly accrual of the costs of organizing a position with chair, desk and other office items

5.16.6.Opportunity Costs

One of the most important factors in a queue are dropouts. Due to impatience or time constraints by the customer it is possible that it may decide to abandon the system, which for the company represents a cost of image quality and timeliness unable to sell and meet the needs of a client. This cost is therefore determined by the abandonment of the system and is defined as:

$$Cop = CoAb$$

Where represents the opportunity cost that a customer leaving the system, and Ab dropouts per node.

5.16.7.Other Costs

Finally, there are other specific costs as the company that can be specified. Generally, these costs depend on clients and servers.

$$Cot = Ctl\ (S + K)$$

Ctl which represents indirect costs per person and other costs related to the service, such as saturation indirect costs of commercial premises such as fines, property damage, wear property, plant and equipment, etc.

5.17. OPTIMIZATION MODEL

From the categories defined above it is possible to build an optimization model in order to minimize service costs and minimize the number of people in queues, and therefore waiting times for customers. With this model we will be developing a support tool for decision-making in business management services aimed at improving efficiency and quality in their care systems.

5.17.1.Definition of Variables

To set the optimization model is necessary to define all variables and parameters in the system interact and affect their performance. Within the configuration of the models and factors that affect these are: the arrival rates and Care servers, their probability distributions, the ability of each of the nodes in the system and the number of servers in the system. Within this list the amount of available servers in the system and capacity are variables, because although other factors affect the objective of minimizing the waiting time, these are values that are out of control of a company. For example, the number of nodes in the network, is a factor that while it may be modified by organizations would only be possible with a long and careful study of the processes in the network are performed, including modification implies the complete transformation of a place of care. These technical and economic difficulties show that the variation of these factors is not feasible in close time horizons.

5.17.2. Objective Functions

Companies within the continuous improvement process are attentive and eager to proposed changes to optimize its operations and focus on the best optimization is to provide the best service at the lowest possible cost.

To quantify this objective, they proceed to form equations as a guide to reach the optimum system solutions.

The first objective function is described by HA Taha in his book Operations Research. In it seeks to minimize the total cost of a queuing system from all expected costs and associated service. This equation are added costs identified other system utilization, jobs and others.

However, although this objective function is functional does not include situations where expected costs are not tangible; If this is the case the result would discount completely the value of Lq, the result is vital in optimizing a queuing system.

$$\min Z = (Cs + Cpt)\sum Sn + (Cu + Ctl)(\sum_n Kn + \sum_n Sn) + Ce\sum_n Lqn + CO\sum_n Abn$$

The ideal and recommended for companies is to conduct an evaluation exercise of factors that enable tangible return costs are not. Despite this, if the last factor is not possible, in addition to the current target feature a unique feature to minimize the waiting time in the queue is derived from the number of people in the system adds.

$$\min Z = \sum_n Lqn$$

This objective function will be used regardless of whether the cost of waiting is tangible or not, and looking on one side to prevent the effect of Lq is eliminated in the optimization; and in the case that the cost expected to be known, seeks to incorporate a greater weight to the number of people in the queue.

5.17.3. Constraints

Restrictions define a logical framework and permissible action in accordance with the needs of a problem. For the case study they should be associated with these variables and outcome objectives.

5.17.4. Constraints associated with Servers

The first of these restrictions see designed to restrict the maximum number of servers within the system. This restriction is related to the maximum allowable cost organizations for hiring employees and/or physical care facilities only allow a certain number of servers.

$$Sn < Stn, \text{ for each n}$$

$$\sum_{n} Sn < \sum_{n} Stn$$

5.17.5. Constraints Associated with Capacity

The capacity constraint defines a framework for assessing to what extent it is possible to expand the capacity of a node.

$$Kn < Ktn, \text{ for each n}$$

$$\sum_{n} Kn < \sum_{n} Ktn$$

5.17.6. Restrictions Associated with Costs

Every business has a balance point where the result of regular exercise to produce a good or service does not represent lost or earnings. By calculating this indicator is possible to identify what the maximum cost that can withstand without incurring losses. This value is comprised of the total sum of the

Payroll servers, charges, fiscal contributions, contributions to security, the expected total cost, the cost of system maintenance, the costs of the job, technology and other costs referenced.

$$Car + Cadm + Cman + Crh + Cas + Cim + Csp < C$$

5.17.7. Final Model

After defining the variables and parameters, build the objective functions and constraints, the optimization model in its compact version is:

sets: N = Number of nodes parameters: λ = Arrival rate μ = Rate care per server **kn**= Capacity at node n.	objectives $$\min Z = (Cs + Cpt)\sum Sn + (Cu + Ctl)(\sum_n Kn + \sum_n Sn) + Ce\sum_n Lqn + C0\sum_n Abn$$ $$\min Z = \sum_n Lqn$$
St= Total number of servers **No**= Number of nodes in the system C = Maximum System Cost variables : **Sn**= Number of servers on the node	restrictions: $Sn < Stn$, for each n $$\sum_n Sn < \sum_n Stn$$ $Kn < Ktn$, for each n $$\sum_n Kn < \sum_n Ktn$$ $$Car + Cadm + Cman + Crh + Cos + Cim + Csp < C$$ N. K. Sn. St. No ε Z λ. μ. ε R C

5.18. MODEL SOLUTION

In order to respond to such problems where it is desired to choose the best combination system, the optimization model poses whose formulation shows objective functions that cannot be assessed by traditional mathematical methods, since now there are no formulas delimit the models developed here.

Such problems are complex to solve because the evaluation of the equations and the uncertain nature of the possible values that can take the objective functions. Only with the simulation of each scenario can be assessed equations and find the optimal values of the variables. The difficulty arises when queuing systems in general are combinatorial problems where the set of solutions is quite large and its evaluation is computationally expensive (NP-hard), therefore it is necessary to use heuristic optimization methods.

A metaheuristic is a procedure for solving high mathematical complexity that under standard procedures at a precise time period prudent. While this method looks very good solutions are not necessarily the optimal solution cannot come, but if the algorithm is well designed, it is possibly very close to the optimal solutions are achieved. While there are very good techniques of this type, it is time to explore other methods using heuristic search optimization that make use of the theory. Therefore, to solve the current model an optimization methodology that involves concepts of decision-making with local search algorithms will be developed.

5.18.1.Evaluation and Selection of Heuristics via Game Theory

The queuing networks as were modeled in this study will be optimized based on the two functions defined objective minimizing total cost (function Z) and minimizing Lq (Y function), each of which has a set of objects which they are called states, and they together form search space.

The complication of this model is that within the sets no elements producing optimality in both objective functions simultaneously. The conflict between the objective functions will make the improvement in function worsen the other, i.e., that the improvement in performance indicator Lq only be possible with increased operating costs of the system.

To explain and address this situation the model is in terms of a game. Suppose an organization with two areas: Quality and Finance. Both in its role seek actions in order to exercise better management and this is reflected in the productivity of the organization to which they belong. If the President of this organization asks at a meeting of both areas on how to make your business better would meet two different answers.

Financier surely making a statement of the constraints of the economy and the effects of having a high payroll in operating costs of care business and the value of the company itself, suggest downsizing or does not increase this resource in order to value the company and reduce the obligations it has.

The quality manager meanwhile discard the above solution and propose increasing the level of care more workers with the aim of generating a quality service that enables the business to expand its market and improve their profits.

For the President is a fact that the quality offered at that time is lower than expected at a company in its class, but also knows the budget constraints and the need for shareholders to increase their profits. Therefore, it proposes to his subordinates think about what is best for the company.

Within each area two different strategies, one of which is to present a plan to action convinced that its proposal will enable the company more useful than the proposal of the other area increasing internal productivity indicators full, or file shall be a plan B that while improving its indicators estimated an edge of the proposal to the other area responsible.

Negotiation raised in this way a game in which every area of the company would have two strategies with a payoff matrix defined by the percentage improvement provided by the area indicators of productivity of the whole company would settle.

Table 5.17 shows the game in matrix form with the strategies and payments (a, b) under the assumptions of the game. If both areas chose to present the plan to, any decisions which will make the improvement provided by each area is zero will be taken, because the two strategies will be completely opposite. If, however, one area decides to take Plan A and another Plan B, the input area with plan A to the efficiency indicators of the company will be an improvement of 10%, while the other will not provide nothing. Finally, if both decide to submit a plan B the contribution that each area has in improving the indicators will surround 5%.

Table 5.17. Set the Objectives

	plan A	Plan B
plan A	(0, 0)	(10, 0)
Plan B	(0, 10)	(5, 5)

If the problem of cost and quality were strictly related to the process of customer service, this game would represent the optimization problem as I was raised in the preceding paragraph. The vector (a, b) improvement would consist of elements function Z and Y respectively function. The concept of optimality is now relative, since there are several solutions with the same results but with different levels of satisfaction. For example, the result (10.0) and (0.10) are optimal solutions with equal benefit. This concept is not new and is known as Pareto efficiency. If something generates or produces benefit, comfort, fruit or interest without harming another, will cause a natural process optimization to reach the optimal point.

Then, an X solution consisting of the elements of the target Z functions and Y is a Pareto optimal if there is no other solution that improve one of the objectives to the values obtained for X without damaging simultaneously one of the other. In general, the solution Pareto to multi-objective optimization problem is not unique and will consist of all solutions with minimal in some function to the values of the other.

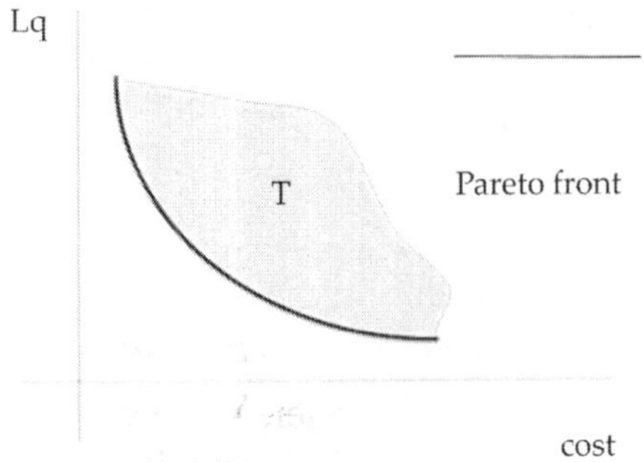

Therefore, while the costs can be lower without affecting the number of waiting in queues are not optimal in the sense Pareto situations. The line marking the border minimum all possible solutions (f) in Figure 40 is called Pareto front, as are all Pareto optimal points within the problem.

However, as shown above, within optimal such an imbalance in the target punishing one in order to achieve better results in the other functions. The goal then it will reach a point of satisfaction for both entities in the game and where both objective functions are met.

To meet this must look for a solution in which no possible benefit more objective without harming the other and at the same time represents a balance between the objectives initially. This solution would be the combination (5,5).

Due to the large number of possible combinations, the search for optimal solutions in the Pareto frontier, be made with a local search method. This algorithm will select within the universe of possibilities nearby possible solutions to Pareto optimality then it will select those that represent the best balance between the objective raised functions.

5.18.2.Algorithm Design

Local search is a technique that comes from artificial intelligence and is based on a structure of adaptive memory that allows we to explore a solution space efficiently through the neighborhood structure. Due to the complexity of searching within the set solutions, we need to take a tour that allows find points near Pareto optimal and from there to assess whether or not the solution.

5.18 3.Initial Setup

A configuration is a set of integer variables of the problem arranged in an array (vector comprising setting the factors in a queue) 122, and the initial setup is from which part the algorithm in order to improve their current state. The initial setup will be a Pareto optimal known. In this case, a server in each of the existing nodes, which although represents the maximum savings, does not account for anything Lq results will be taken.

5.18.4.Generation Neighborhood

A neighbor of an X configuration is a configuration X 'obtained from X. For the problem of tails, a neighbor is generated in a unit increasing the number of servers on the node i consecutively increasing capacity.

5.18.5.Selecting the Best Neighbor

In generating the neighborhood another concept used in operations research and dynamic optimization is introduced. It shows that it is possible to break a problem into stages and from the optimum found in one of them it is possible to find an optimum in the following stages to form an optimal solution for all problems.

If we seek within the universe of configuration possibilities, there are many possibilities, but many of them would not present improvement in the desired result. With the optimization stages seeks to limit the neighborhood using the knowledge system has to add intelligence in choosing possible solutions within the Pareto frontier.

According to that seen in chapter analysis results, the impact of the increase of server performance measures is more evident if the increase occurs in the first nodes, this means that from optimal server on the first node it is possible to find optimal in the following nodes that depend on the optimal configuration of an already known node.

In this chapter also it evidenced how variation server capabilities and have a greater impact on Lq is increased when the top levels; then presumably impact queues variation from one to two servers is much more significant than the change in quantity when more servers.

This leads to the conclusion that as progress is made in the search for optimal solutions in the Pareto sense, will reach points where would improve starts to become less noticeable; and from there, regardless of the number of additional iterations, the Pareto frontier will already be defined.

5.18.6.Assessment of Neighbors

A neighbor is generated from an X position cannot be selected again unless it is dominant over previous configurations. The Pareto optimality criterion said X is a vector composed of the objective functions solution (z, y) if and only if no other vector that dominates.

A vector x = (x1; x2) Dominates another x '= (x'1; x '2) If and only if x is partially less than x

$$\forall i \in \{1,2\} \quad xi \leq xi' \wedge \exists i \in \{1,2\} \quad !xi < xi'$$

With this approach the search space will travel to find dominant points leading to the Pareto optimal, or at least close to it to the extent possible to select a balance between the objective functions results.

After finding the Pareto frontier is likely the dilemma of which of these solutions represents a balance between functions. Where is then raised back to the game: what solution would simultaneously both areas (bonds and quality) were satisfied? For this example would be the point (5,5), but further expanding the spectrum of possibilities and oriented towards the rational behavior of an individual will assume that the balance between the interests of the areas will only be possible as a solution represent equivalent benefits. For that purpose and because of the incompatibility of the units between the functions are used to target method, but was conceived in another context, it is useful for solving this game and elasticity.

In economy, the elasticity is the ratio formed between the proportional change of a variable with respect to the proportional change. Mathematically speaking, the elasticity is proportional change of a dependent variable (cost function) relatively to another independent variable (Lq):

$$E = \frac{\frac{\Delta y}{Y}}{\frac{\Delta z}{Z}}$$

Focusing on the concept and the model formula, if the indicator is greater or less than unity represents the variation of the target Pareto optimal solution to another tends to benefit more functions to an area; and if on the other hand, the elasticity is unitary we have a balance in the benefits that represents the optimum satisfaction sought.

Additional to this approach, it may directives service center believe that the weight of one of the two functions is more important than the other, which is added to the model one α parameter, which assign a weight to the functions aim to choose the balance between them depending on the selected criteria.

Accordingly, the evaluation formula of Pareto optimal X (z, y) versus another is:

$$E = \frac{\Delta y Z a}{\Delta z Y(1 - a)}$$

The optimum model will be generated by the solution whose change possesses a closer one elasticity

5.18.7.Stopping Criteria

The algorithm can search and continue indefinitely neighbors picking up walking the whole problem. However, in large and with many possible solutions problems this task may become slow and produce results in fairly large times. In order to enable convergence of results in a timely manner and with a good approximation of what would be optimal stopping criterion is established.

As discussed previously, the variation of servers and capacity have a greater impact if there are fewer servers and the system capacity is limited; then, with increasing the number values or iterate among neighbors to find solutions, the results will lose increasing impact on performance indicators.

Therefore, the stop criterion is determined by successive attempts to improve without success. If in subsequent iterations after finding a local optimum, no improvement occurs and this continues for at least K/10 (number of possible skill levels within problem) iterations, the algorithm stops searching process, since this at close to optimum, or at least one better than their predecessors

5.18.8.Functioning

In short, the design algorithm developed here consists of the following stages:

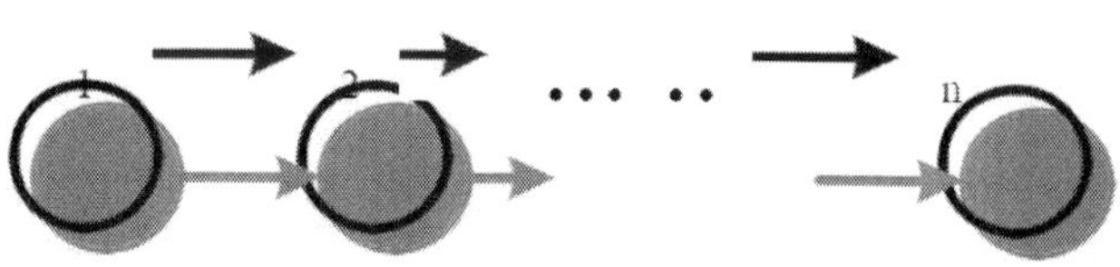

Figure 5.27. Fractionation of network nodes.

(Figure 5.27) Start Said process is completed in each of the stages are represented by the nodes.

generate the
neighbor (S + i)
Calculate
objective functions
determine front
Pareto
Evaluate x 'is
better balance
X
Yes
Update x
with x '
Meets
Criteria
Stop
Yes
End

5.18.9.Evaluation

Within this type of heuristics that use the Pareto frontier as optimization criteria have been developed or errors scattering measurements in order to evaluate how rough is the algorithm in obtaining the points that meet the criteria Pareto.

One such measure was developed by Jason Schott in order to measure the variance of the distance of each member of the Pareto optimal set with respect to its nearest neighbor.

$$D = \sqrt{\frac{1}{n-1}\sum_{i}(d - di)^2}$$

Where $di = min\,(I\,(z - zj)\,I + I\,(yi - yj)\,I), i, j = 1,2,...,n$.; d is the average of all di n is the number of elements of the set of Pareto obtained so far. If means that the algorithm has found the ideal distribution of vectors dominated.

5.18.10. Example

To make a detail of how the algorithm works together with the results shown, it seeks to develop an illustrative example

The model input parameters are determined by the maximum level of the experiment of this study represent the constraints of the optimization model, in order to equate the results and glimpse the Pareto frontier calculated in relation to the universe of possibilities problem. Parameters, and other costs specified below:

The variance between the points of the Pareto border according the above formula is:

D = 0.16 (average for each node)

This value of D says it is a good approximation to the optimization algorithm.

Figure 42 presents a table showing the results of the algorithm, there exposed table classified by the number of servers the results of objective functions and also the identified Pareto chart border with the optimum (red spot) selected.

	Nodo 1		Nodo 2		Nodo 3	
1.00	14,563,160.94	9.24	14,158,972.59	8.77	14,339,780.14	8.86
2.00	14,029,246.66	7.69	13,707,423.04	7.49	13,867,024.88	7.45
3.00	13,212,214.62	3.58	13,379,171.16	4.35	13,190,426.46	3.38
4.00	13,806,654.76	0.74	13,807,534.76	0.84	13,802,930.34	0.33
5.00	14,504,552.87	0.51	14,503,250.25	0.36	14,503,397.42	0.38
6.00	15,200,479.78	0.05	15,200,578.40	0.06	15,200,411.88	0.05
7.00	15,900,258.78	0.03	15,900,009.77	0.00	15,900,162.64	0.02
8.00	16,600,009.92	0.00	16,600,000.00	-	16,600,262.60	0.03
9.00	17,300,000.00	-	17,300,000.00	-	17,300,025.05	0.00
10.00	18,000,000.00	-	18,000,000.00	-	18,600,000.00	

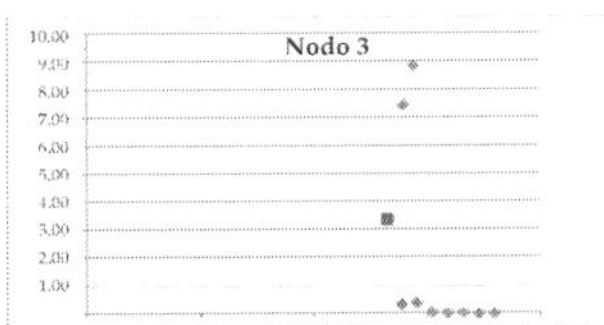

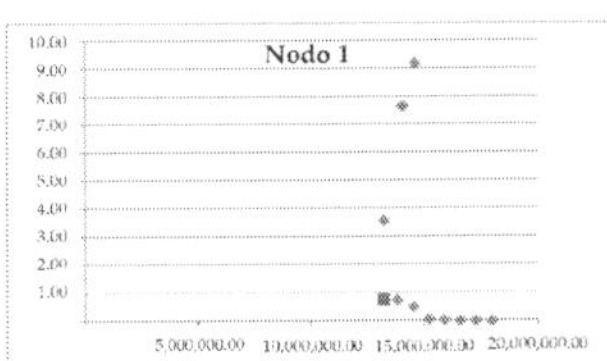

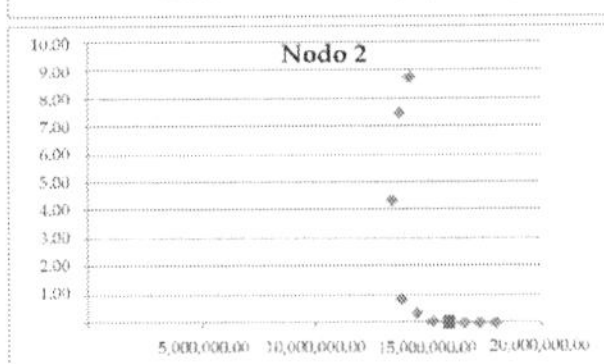

Figure 5.28. Border Pareto by nodes in a network queue.

5.18.11. Application Tool

Development and analysis underlying this study is based on the theory of diverse knowledge of operations research and engineering that cannot be fully known or dominated by one person in charge of managing a queuing system. So that both the analysis and solution methodology are useful for decision-making at the enterprise level, a tool that simplifies is prepared and meets the assumptions and procedures in a simple application, easy to understand and easy to use. Through this application the user can optimize your system queue according to the parameters, variables and your company own characteristics, provided that they can be coupled to the scope of this study.

5.18.12. Information Utilized

The tool involves the use of four files and two in Excel Arena. The first file in Arena is the interface of the application as such; and the remaining three networks are one, two and three nodes to simulate the input information (nodes 1, 2 and 3). Moreover, Microsoft Excel files are to print the results of optimization.

Performance requirements

The characteristics necessary to consider the application to run successfully are:

- Install Rockwell Arena simulation software version 10 or higher, preferably with commercial license. However, the academic version can be useful if the number of entities within the system is less than 150.
- Configure the system with decimal point (e.g., 3.05)
- Should have him eat, go to Control Panel/Regional and Language Options/Customize Format.
- Checking the activation of the complementary database, spreadsheet models and Arena, among others, in bookstores and references Excel

5.18.13. Tool Operation

The application is based on a process of feeding information, internal calculation results and presenting them to the user. This procedure is shown in Figure 5.29

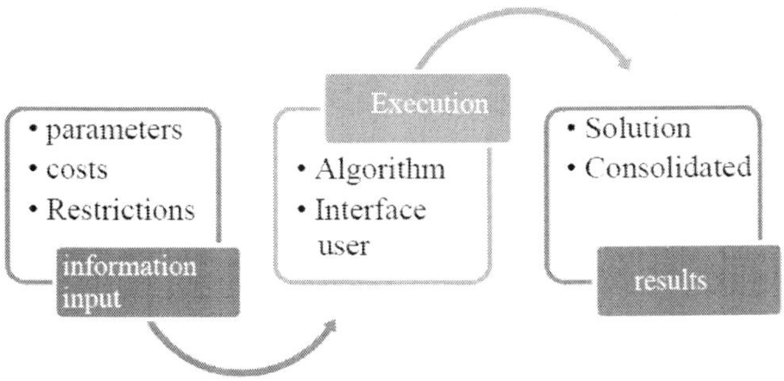

Figure 5.29. Outline of the tool

5.19. TOOL DESIGN

The application has a friendly interface and easy to understand allowing the user to interact quickly to optimize your system once known input information. With the idea of simplifying and harmonizing the application a series of windows, formats and forms that are useful for entering inputs and outputs understanding were designed.

5.19.1.Input Information

The tool user must supply certain inputs to emulate the queuing system. For the proper functioning of the algorithm and design issues, such information is divided into three separate tabs: Parameters, Costs and Model. Next, we describe each one of them.

5.19.2.Parameters

In this tab we must enter the parameters of the optimization model. Some of them are limited by the levels of the experimental design of this study and others are free entry to depend on each company, such as arrival rates and service.

The costs of a system should be introduced in a form with the same units and taking into account monthly.

5.19.4.Model

It is this tab the parameters that define the restrictions and the importance of the objective functions are introduced. Here is the optimize button to be pressed once all the information has been registered in the system. If we have not registered all information and polish the Optimize button. It is necessary for the proper functioning model introduce missing values or unknown as zeros in the model, if not so processing errors can occur impeding the solution model.

It is important to note that the capacity determines the number of people who can attend at each node, therefore, that parameter is defined by a positive integer variable that must be multiple of 10 according to the levels of the experiment. To introduce a number other than multiples of 10 model area adjustment, asked the values entered since this course is essential for the proper functioning of the algorithm.

5.19.5.Output Information

The output is the optimal server configuration and capacity of each node, along with detailed information of the target and dominant solutions according to the optimization criterion functions.

The results are saved and the file key information Resultados.xlsm model and solution thereof will be found. This file will automatically open once we finish come the simulation model

CHAPTER 6

STOCHASTIC MODELING AND SIMULATION

CONTENTS

This chapter introduces the notion of counting processes to present the M/G/1 system from the perspective of stochastic processes. Some quantities of interest for Queuing deduced and certain variations of M/G/1 system is presented. Besides the problem of M/G/k system highlighting its open condition problem arises. Developments presented in this chapter are based on developments Sheldon Ross (1993). Finally the inclusion of a limit number of customers in the system and its effect on some variables of interest is discussed.

6.1. COUNTING PROCESSES

A stochastic process $E[N(t)] = \lambda t$ counting process is said if N (t) represents the number of events occurring until time t. A counting process must satisfy:

- $N(t) \geq 0$.
- N (t) is an integer value.
- If s <t, then N (s) <N (t).
- If s <t, N (t) – N (s) represents the number of events in the range (s, t).

A Poisson process $E[N(t)] = \lambda t$ justly λ It is a counting process that satisfies the following conditions:

- N (0) = 0.
- It is independent increments.
- The number of events that occur in the long interval t has a Poisson distribution with mean λ t, i.e.:

$$P\{N(t+s) - N(s) = n\} = e^{-\lambda t}\frac{(\lambda t)^n}{n!}, \qquad n = 0,1,...$$

In addition it holds that $E[N(t)] = \lambda t$.

We define as the inter-arrival time sequence to the sequence $\{T_n, n = 1,2,...\}$, where T_n represents the elapsed time between the event n-1 and the nth time. Distribution behaves like:

$$P\{T_1 > t\} = P\{N(t) = 0\} = e^{-\lambda t}$$

$$P\{T_2 > t\} = P\{T_2 > t / T_1 = s\} = P\{N(t+s) - N(s) = 0\} = e^{-\lambda t}$$

T_n, N = 1,2,..., are exponential half $1/\lambda$.

The waiting time until the nth event is given by:

$$S_n = \sum_{i=1}^{n} T_i, \qquad n \geq 1$$

which it has a Gamma distribution with parameters n λ, i.e.:

$$f_{S_n}(t) = \lambda e^{-\lambda t} \frac{(\lambda t)^{n-1}}{(n-1)}, \qquad t \geq 0.$$

6.1.1. Compound Poisson Process

A stochastic process $\{N(t), t \geq 0\}$ a Poisson process if said compound can be represented by:

$$X(t) = \sum_{i=1}^{N(t)} Y_i, \qquad t \geq 0$$

where $\{N(t), t \geq 0\}$ It is a Poisson process and $\{Y_n, n \geq 0\}$ is a family of random variables which are also independent of $\{N(t), t \geq 0\}$.

6.2. THEORY OF TAILS

6.2.1. Definition of the Problem

The problem is to model the dynamic behavior of some variables of interest of a customer service system. The quantities of interest are considered in the model are:

S: number of servers.

n: number of customers in the system.

N: maximum number of customers in the system.

L: average number of customers in the system.

L_Q: Average number of customers waiting in line.

W: average time a customer in the system.

W_Q: Average time a customer waiting in the system.

In addition we can define the average arrival rate of customers λ_a by:

$$\lambda_a = \lim_{t\to\infty} \frac{N(t)}{t}$$

Then we can raise the following formulas:

$$L = \lambda_a W$$

$$L_Q = \lambda_a W_Q$$

6.2.2. The M/G/1 System

Notation is given as follows.

M: Poisson arrivals rightly λ.

G: General distribution service.

1: a server.

In this model, customers will be served in order of arrival FIFO (first input, first output). In addition there will be no limitations on the number of customers that the system supports (customers can be served infinite). Define the legacy system to a customer at a time t as the sum of the remaining time of service all customers in the system at that moment. V denotes the average working time in the system. It's time service W_Q^* the time a customer must wait in line. So:

$$V = \lambda_a E[SW_Q^* + \int_0^S (S - x)dx]$$

$$V = \lambda_a E[SW_Q^*] + \frac{\lambda_a}{2} E[S^2]$$

Overall, the service time is independent of time in line, i.e., :

$$V = \lambda_a E[S]W_Q + \frac{\lambda_a}{2} E[S^2]$$

As the system is FIFO holds that $W_Q = V$, so that:

$$W_Q = \frac{\lambda E[S^2]}{2(1 - \lambda E[S]\}},$$

where E [S] and $E[S^2]$ They are the first two moments of the distribution service. Then it is possible to obtain the following amounts of interest:

$$L_Q = \lambda W_Q = \frac{\lambda^2 E[S^2]}{2(1-\lambda E[S])}$$

$$W = W_Q + E[S] = \frac{\lambda E[S^2]}{2(1-\lambda E[S])} + E[S]$$

$$L = \lambda W = \frac{\lambda^2 E[S^2]}{2(1-\lambda E[S])} + \lambda E[S]$$

Here are the next few important observations:

i) From Theory to renewal processes, if a server is always busy, the reason for starting $1/E[S]$ It will be greater than the ratio of arrival λ. So λ E [S] <1, implying that the quantities L, L_Q, W and W_Q They are finite.

ii) As Var [S] = $E[S^2] - E[S]^2$ If the average service time is constant, any increase in the variance is reflected as an increase in L, L_Q, W and W_Q.

iii) There are alternative ways to get W_Q.

6.2.3. Leisure Time and Work of the Server Application

A system will alternate between working time (when there is at least one client and in the system) and leisure time (idle server). We note by I_n Y B_n the lengths of the n-th periods of work and leisure respectively $(n \geq 1)$. The proportion of time the server is idle is:

$$P_0 = \lim_{n\to\infty} \frac{I_1 + . + I_n}{I_1 + ... + I_n + B_1 + ... + B_n}$$

where $I_1 + . + I_n$ Y $B_1 + ... + B_n$ are random variables. Applying the law of large numbers:

$$P_0 = \frac{E[I]}{E[I] + E[B]},$$

with I and B random variables representing leisure time and work. As I represents the time from when a customer leaves the system and no one left, until the next one, I is rightly exponentially distributed λ and we have:

$$E[I] = \frac{1}{\lambda}$$

Since $P_0 = 1 - \lambda E[S]$, we have:

$$1-\lambda E[S]=\frac{\frac{1}{\lambda}}{\frac{1}{\lambda}+E[B]}$$

$$\therefore E[B]=\frac{E[S]}{1-\lambda E[S]}$$

Be a_n the proportion of customers who are n customers in the system when they arrive. C is the number of clients served in a work period. Then each E [C] customers one will find only the empty set, i.e.:

$$a_0=\frac{1}{E[C]}$$

But $a_0=P_0=1-\lambda E[S]$ and so:

$$E[C]=\frac{1}{1-\lambda E[S]}.$$

6.3. VARIATIONS MODEL M/G/1

6.3.1. M/G/1 with Random Arrival Batch Size

If customers arrive at the same time at the system, this means that the number of customers can be increased in a time unit in amounts greater than one, known as the amount batch size input. N we note by the random variable representing the size of the batch, i.e.:

$$P\{N=j\}=\alpha_j$$

Formula work in the system remains:

$$V=\lambda E[N] E(S)W_Q+\frac{E(S^2)}{2}]$$

However, the waiting time is increased by adding a time related to the time system the other customers who are in the same batch, i.e.:

$$W_Q=V+E[W_B]$$

where W_B is the timeout induced batch. The proportion of customers who may come from a batch size j is proportional to the size of the batch and its probability α_j, i.e.: Customer selectable ratio is

$$\frac{j\alpha_j}{\sum_j j\alpha_j}$$

If our client is in the i-th position in your batch, we must wait for i-1 customers before it is served, i.e.:

$$E\left[{W_B}/{Batch_{size} = j}\right] = \sum_{i=1}^{j}(i-1)E(S)\frac{1}{j} = \frac{j-1}{2}E[S]$$

$$\therefore E[W_B] = E\left[{W_B}/{Batch_{size} = j}\right].$$

Select proportion of customers. Batch size j

$$E[W_B] = \frac{E[S]}{2E[N]}\sum_j (j-1)j\alpha_j$$

$$E[W_B] = \frac{E[S]}{2E[N]}\left(E[N^2] - E[N]\right)$$

so:

$$W_Q = \frac{E[S]\left(E[N^2]-E[N]\right)/2E[N] + \lambda E[N]E[S^2]/2}{1-\lambda E[N]E[S]}$$

Observations:

- W_Q is finite ssi $\lambda E[N] < 1/E[S]$.
- To E [N] fixed, W_Q Var is proportional to [N].
- From W_Q We can deduct amounts of interest L, L_Q and W considering:

$$W = W_Q + E[S]$$

$$L = \lambda_Q W = \lambda E[N]W$$

$$L_Q = \lambda E[N]W_Q$$

6.3.2. M/G/1 with Priority Queues

Suppose there are 2 types of customers: the first one waiting in the system to customers who have type 1 arrived before him and only waits for a client type 2 if this was in service on arrival. Type 2 client should expect that all customers are cared for type 1 and must wait for all type 2 who have come before him. Poisson arrivals are distributed with reasons λ_1 Y λ_2 and the service has distributions G_1 Y G_2. We note by W_Q^i the average time in the client queue type i, i = 1,2. Among clients of the same type as the priority is FIFO, the model behaves like a M/G/1 and their distributions obey the following relationships:

$$\lambda = \lambda_1 + \lambda_2$$

$$G(x) = \frac{\lambda_1}{\lambda} G_1(x) + \frac{\lambda_2}{\lambda} G_2(x)$$

It is known that the work is given by:

$$V = \frac{\lambda E[S^2]}{2(1 - \lambda E[S])}$$

then $\lambda = \lambda_1 + \lambda_2$:

$$V = \frac{\lambda_1 E[S_1^{\,2}] + \lambda_2 E[S_2^{\,2}]}{2(1 - \lambda_1 E[S_1] - \lambda_2 E[S_2])}$$

we note by V^i the average amount of work i type:

$$V^i = \lambda_i E[S_i] W_Q^i + \frac{\lambda_i E[S_i^2]}{2}$$

Let us define the following amounts:

$$V_Q^i = \lambda_i E[S_i] W_Q^i$$

$$V_S^i = \frac{\lambda_i E[S_i^2]}{2}$$

We interpret the quantities V_Q^i and V_S^i as the average amount of work type 1 in the queue and the average amount of work service type 1, respectively. Then we can ask the following relationship:

$$W_Q^1 = \frac{\lambda_1 E[S_1^{\,2}] + \lambda_2 E[S_2^{\,2}]}{2(1 - \lambda_1 E[S_1])} \qquad (*)$$

As $V = V^1 + V^2$ so:

$$\frac{\lambda_1 E[S_1^{\,2}] + \lambda_2 E[S_2^{\,2}]}{2(1 - \lambda_1 E[S_1] - \lambda_2 E[S_2])} = V_Q^1 + V_S^1 + V_Q^2 + V_S^2$$

$$= W_Q^1 + V_Q^2$$

$$= W_Q^1 + \lambda_2 E[S_2] W_Q^2$$

then using the previous we have:

$$W_Q^2 = \frac{\lambda_1 E[S_1^2] + \lambda_2 E[S_2^2]}{2(1-\lambda_1 E[S_1] - \lambda_2 E[S_2])(1-\lambda_1 E[S_1])}$$

Observations:

- W_Q^1 E4S a finite amount ssi $\lambda_1 E[S_1] < 1$.
- W_Q^2 is finite ssi $\lambda_1 E[S_1] + \lambda_2 E[S_2] < 1$.

If there are n types of customers, we will make the same amounts above analysis looking V^j, J = 1,... n. The amount of work on the client system type 1,..., j only depends on its priority on customer type j + 1,..., n. Then we can define two general classes I and II priority such that:

$\lambda_I = \lambda_1 + \ldots + \lambda_j$,

$\lambda_I = \lambda_{j+1} + \ldots + \lambda_n$

According to the above, the expectations of the moments of time associated services are given by:

$$E[S_I] = \sum_{i=1}^{j} \frac{\lambda_i}{\lambda_I} E[S_i],$$

$$E[S_I^2] = \sum_{i=1}^{j} \frac{\lambda_i}{\lambda_{II}} E[S_i^2],$$

$$E[S_{II}^2] = \sum_{i=j+1}^{n} \frac{\lambda_i}{\lambda_{II}} E[S_i^2]$$

Further work V^I is:

$$V^I = \frac{\lambda_I E[S_I^2] + \lambda_I \lambda_{II} E[S_I] E[S_{II}^2]}{2(1-\lambda_I E[S_I])}$$

where $V^I = V^1 + \ldots + V^j$ for each j = 1,..., n, it is generating it from (**) can deduce the following relationship:

$$W_Q^i = \frac{\lambda_1 E[S_1^2] + \ldots + \lambda_n E[S_n^2]}{2\prod_{j=i-1}^{i}(1-\lambda_1 E[S_1] - \ldots - \lambda_j E[S_j])}, \quad i = 1,\ldots, n$$

6.3.3. The M/G/k System

In a system like this customers arrive with Poisson distribution ratio λ and they are served by one of the k existing servers, each of which has a distribution service G. As in the M/G/1 system, the work of the system is given by the expression:

$$V = \lambda_a E[S]W_Q + \frac{\lambda_a}{2} E[S^2] \quad (3)$$

If there are k servers, the work will be done at a ratio k per unit time. This means that before an arrival, work on the system is given by the:

$$V = kW_Q + E[R] \quad (4)$$

where R is the sum of the remaining service times of arrival. If E [R] could be calculated, we could clear W_Q using equations (3) and (4). However, estimate E [R] is a problem for an exact answer is so is not known that can be calculated W_Q. Solutions to the problem are given based on approaches that work for some cases, such as the approximation of Nozaki and Ross, from which it is possible to approximate W_Q by:

$$W_Q \approx \frac{\lambda^k E[S^2](E[S])^{k-1}}{2(k-1)!(k-\lambda E[S])^2 \left[\sum_{n=0}^{k-1} \frac{(\lambda E[S])^n}{n!} + \frac{(\lambda E[S])^k}{(k-1)!(k-\lambda E[S])} \right]}$$

This is a good approximation when the service distribution approaches a Gamma or Exponential, but in another cases their behavior is not so good, so this issue remains open.

6.3.4. System with Finite Capacity

The analysis assumed that the system has no limitations on the number of clients that can be in it. But the real systems if they have these limitations so it is of interest to include this situation in the model. We note by N the maximum number of customers in the system. So P_n shall mean the probability of n customers have in the system. It is possible to make the following balance equations:

$$\begin{array}{ll} 0 & \lambda P_0 = \mu P_1 \\ 1 \le n \le N-1 & (\lambda+\mu)P_n = \lambda P_{n-1} + \mu P_{n-1} \\ N & \mu P_N = \lambda P_{N-1} \end{array}$$

To leave an arrival state 0 must occur, so the reason is λP_0. On the contrary, to enter the state 0 must produce output being in state 1, so the reason is μP_1. To leave a state n, with $1 \le n \le N-1$, Must produce an output or arrival being in n, so the reason is $(\lambda+\mu)P_n$. On the contrary, to enter a producible by an arrival being at n-1 or with an outlet being in n + 1, so the reason is $\lambda P_{n-1}+\mu P_{n-1}$. Being in the state of limit number N of clients is only possible to leave the system via an outlet, so the reason is μP_N. On the contrary, to enter the state N can only be done from the state N-1 by an arrival, so the reason is λP_{N-1}. Solving the following system of equations is possible to obtain some interesting results:

$$P_1=\left(\frac{\lambda}{\mu}\right)P_0,$$

$$P_{n+1}=\frac{\lambda}{\mu}P_n+\left(P_n-\frac{\lambda}{\mu}P_{n-1}\right), \quad 1\le n\le N-1$$

$$P_N=\left(\frac{\lambda}{\mu}\right)P_{N-1}.$$

If we solve in terms of P_0. We can get:

$$P_1=\left(\frac{\lambda}{\mu}\right)P_0$$

$$P_2=\frac{\lambda}{\mu}P_1+\left(P_1-\frac{\lambda}{\mu}P_0\right)=\left(\frac{\lambda}{\mu}\right)^2P_0$$

$$\vdots$$

$$P_{N-1}=\frac{\lambda}{\mu}P_{N-2}+\left(P_{N-2}-\frac{\lambda}{\mu}P_{N-3}\right)=\left(\frac{\lambda}{\mu}\right)^{N-1}P_0$$

$$P_N=\frac{\lambda}{\mu}P_{N-1}=\left(\frac{\lambda}{\mu}\right)^{N}P_0$$

If it is satisfied that $\sum_{n=0}^{N}P_n=1$ we can get:

$$P_0 \sum_{n=0}^{N} \left(\frac{\lambda}{\mu} \right)^n = 1$$

$$P_0 = \frac{1 - \lambda/\mu}{1 - \left(\lambda/\mu \right)^{N+1}}$$

and so:

$$P_n = \frac{\left(\lambda/\mu \right)^n \left(1 - \lambda/\mu \right)}{1 - \left(\lambda/\mu \right)^{N+1}}, \qquad n = 0,1,\ldots, N$$

Using the above result, the amount L can be expressed directly as:

$$L = \sum_{n=0}^{N} nP_n$$

$$L = \frac{\lambda \left[1 + N\left(\lambda/\mu \right)^{N+1} - (N+1)\left(\lambda/\mu \right)^{N} \right]}{(\mu - \lambda)\left(1 - \left(\lambda/\mu \right)^{N+1} \right)}$$

APPENDIXS

APPENDIX A

In this section we will see the definition and some properties of a stochastic process that has been used in the course of these notes. This process, which we have called Process House of Cards, has state space the set of natural numbers N and is defined as follows. Consider an increasing sequence [0; 1], such that verifies

$$\lim_{k \to \infty} a_k = 1$$

Family define chains $\{\{W_n^m : n \geq m\} : m \in Z\}$ and n > m where $W_n^m := 0$

is a sequence of independent random variables uniformly distributed in the interval [0; 1]. Thus, W_n^m denotes the position at time n of the string starting at time m at the origin.

A.0.1. Observation. The name House of Cards given to the process reflects the fact that once the chain is in the state n can in the next instant jump to state n + 1 or can return to the initial state 0, as when a home is armed with letters adding one to one new letter at each step.

Given x; and 2 N two possible states of the process, see below as are the transition probabilities of this process. We divided the study in three cases:

We know that the process can jump from state to state x =x + 1 or returning to the source, so it is clear that if

$$W_n^m := (W_{n-1}^m + 1)1_{\{U_{n<aW_{n-1}^m}\}}$$

then:

$$\rho(x, y) = \mathrm{P}(W_n^m = y|W_{n-1}^m = x) = 0$$

Suppose now that x + y = 1, then in this case, because the definition of the process and the distribution of the random variables U_n we have to:

$$\rho(x, x+1) = \mathrm{P}(W_n^m = x+1|W_{n-1}^m = x) = P(U_n < a_x) = a_x$$

3. Finally,

$$p(x, 0) = P(W_n^m = 0|W_{n-1}^m = x) = P(U_n > a_x) = 1 - a_x$$

So we conclude that:

$$p(x, y) = \begin{cases} a_x & if \;\; y = x+1 \\ 1 - a_x & if \;\; y = 0 \\ 0 & in\ other\ case \end{cases}$$

A.0.2. Observation. Note that the fact that implies that:

$$W_n^m \geq W_n^k \quad \forall m < k \leq n$$

$$W_n^m = 0$$

$$W_n^k = 0$$

$$\{\{W_n^m : n \geq m\}$$

therefore $W_n^m = 0$ implies that $W_n^k = 0$ for all $\{\{W_n^m : n \geq m\}$. It then holds that:

The following results describe the process behavior based on their abilities probability transition.

Lemma A.0.3. It is valid that $Q(Q(x, y))_{x,y \in N}$ is a Markov chain with values in N with transition matrix

$$Q(Q(x, y))_{x,y \in N}$$

such that:

$$Q(x,y) = \begin{cases} a_x & if \;\; y = x + 1 \\ 1 - a_x & if \;\; y = 0 \\ 0 & in\ other\ case \end{cases}$$

So:

1. Chain is recurrent non-positive $\leftrightarrow \sum_{n\geq 0} \prod_{k=0}^{\infty} a_{k=\infty}$

2. Chain is transient. $\leftrightarrow \prod_{k=0}^{\infty} a_k > \infty$

APPENDIX B

Generating functions:

The method of generating functions is important in the study of stochastic processes with discrete state space tool. This appendix contains some results concerning meet the generating functions that are useful for the development of Chapter 4 presented in this paper.

B.0.4 Definition. Consider the sequence of real numbers 0,1,.... The power series:

$$F(s) = \sum_{i=1}^{\infty} \varphi_i s^i$$

It can be considered as the transformation that takes succession

$\{\varphi_i\}_{i\in N}$

In the function F (S). If this power series converges in some interval

$-k<s<k$

then the function F (s) is called the generating function of $\{\varphi_i\}_{i\in N}$ succession.

Now consider the stochastic process {Xt} and the sequence of probabilities

$$P_{x(t)} := P(X_t = x)$$

$$F(s,t) = \sum_{x=0}^{\infty} P_x(t) s^x$$

the generating function of the probability $P_x(t)$ with x = 0,1,...Note that as the probabilities Px(T) are bounded for all x and for all ty$^{P/}$

$$\sum_{x=0}^{\infty} P_x(t) = 1$$

for all t , 0, comparing B.2 series with the geometric series we see B.2 converges uniformly with t, at least for

$|s| < 1$

B.1. Some theorems and properties.

Theorem B.1.1. The expected value of the variable is given by:

$$E(X_t) = \frac{\partial F(s,t)}{\partial s}]s = 1$$

Theorem B.1.2. If

$$E(X_t^2) = [\frac{\partial^2 F(s,t)}{\partial s^2} + \frac{\partial F(s,t)}{\partial s}]_{s=1}$$

then $x = 0$ *and definition of variance must then:*

$$var(X_t) = [\frac{\partial^2 F(s,t)}{\partial s^2} + \frac{\partial F(s,t)}{\partial s} - (\frac{\partial F(s,t)}{\partial s})^2]_{s=1}$$

Theorem B.1.3. Let $X_1(t)$ and $X_2(t)$ two independent random variables with distributions $P_x(t)$ and $Q_x(t)$ and generating functions

$F_1(s,t)$

and

$F_2(s,t)$

respectively. Let

$Y(t) = X_1(t) + X_2(t)$

is G(s, t) the generating function of the probability

$$R_y(t) := P(Y(t) = y)$$

Then G(s, t) is given by:

$$G(s,t) = F_1(s,t)F_2(s,t)$$

Finally we present below a result that is useful in many problems using generating functions.

If F (s, t) is analytic in s = 0, then we can write F as the Maclaurin series:

$$F(s,t) = F(0,t) + \sum_{x=1}^{\infty} \frac{F^{(x)}(0,t)}{x!} s^x$$

$$F(s,t) = F(0,t) + \sum_{x=1}^{\infty} \frac{F^{(x)}(0,t)}{x!} s^x$$

Then compared with $x=0$ series

$$F(s,t) = \sum_{x=0}^{\infty} P_x(t) s^x$$

$$P_0(t) = F(0,t)$$

$$P_x(t) = \frac{F^x(0,t)}{x!} \qquad x = 1,2, \ldots$$

APPENDIX C. RESULT QUEUE LQ NETWORKS

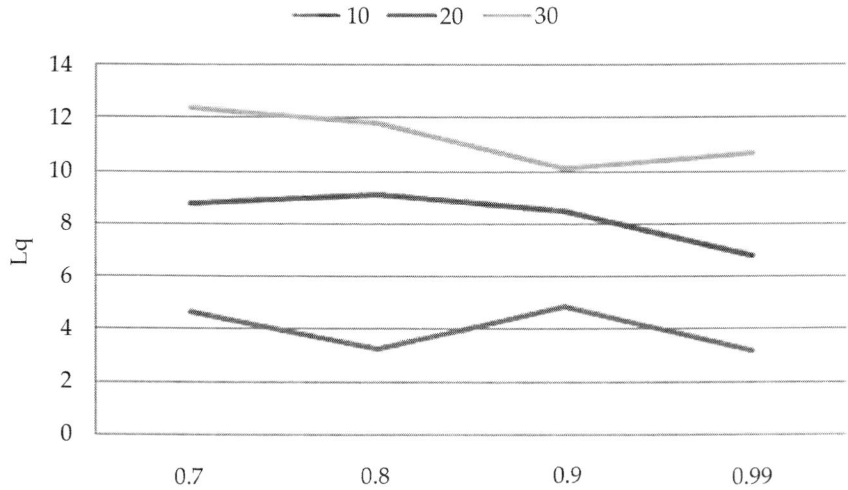

Figure C.1: Results Lq according to the capacity at the second node under the Hyper-Erlang distribution.

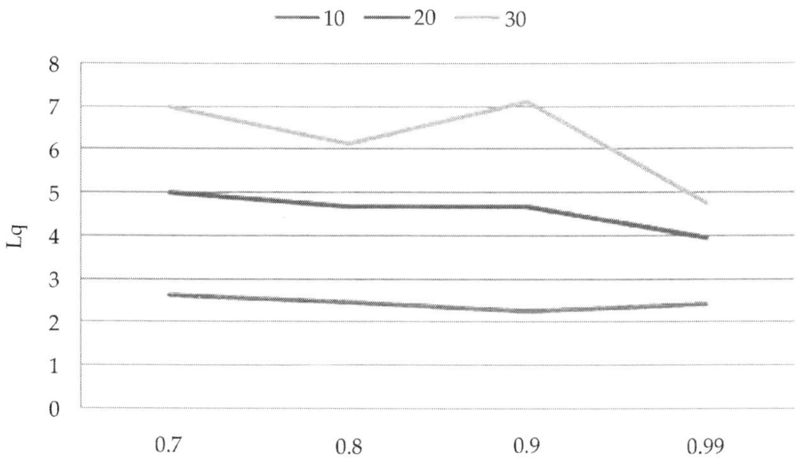

Figure C.2: Results Lq according to the capacity at the third node under the Hyper-Erlang distribution.

APPENDIX D. RESULTS WQ QUEUING NETWORKS.

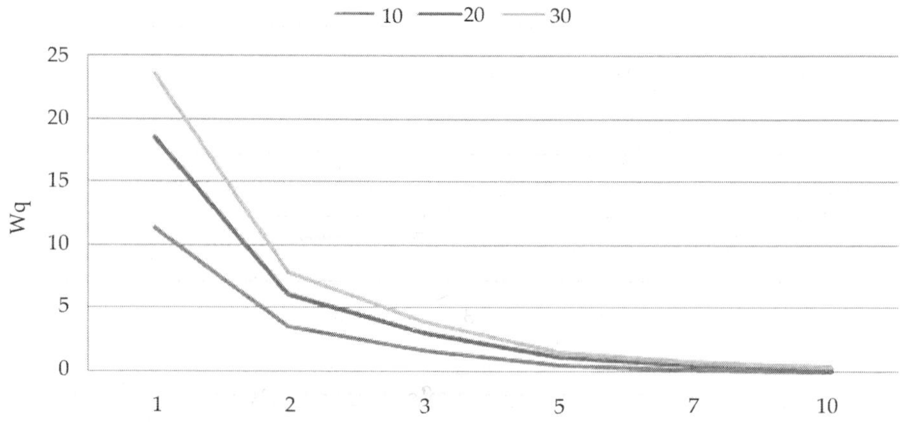

Figure D.1: Results of Wq by the number of servers and capacity at the second node under the Exponential.

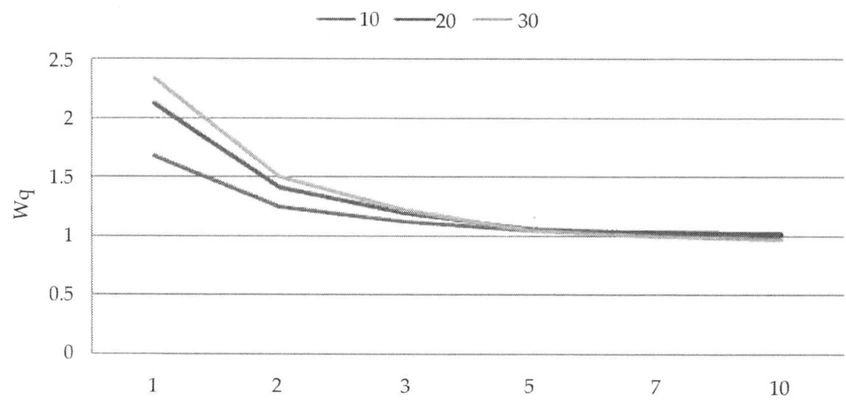

Figure D.2: Results of Wq by the number of servers and capacity in the third node under the Hyper-Erlang distribution

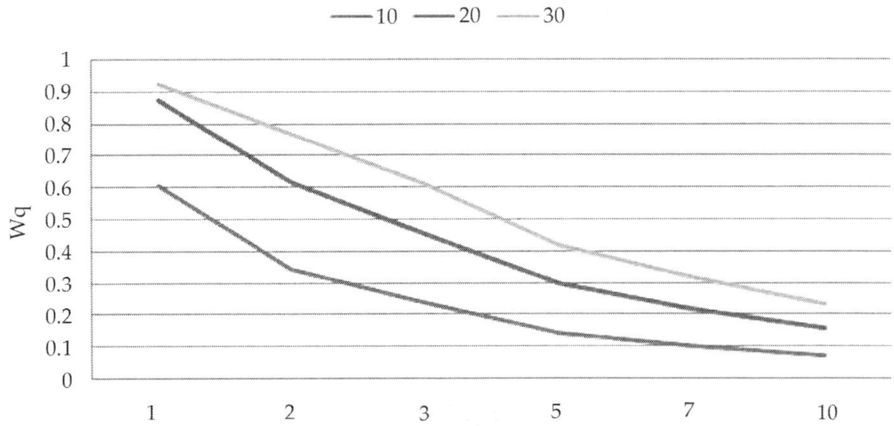

Figure D.3: Results Lq as the number of servers and capacity in the third node under Hyper-Erlang distribution.

CONCLUSIONS

- The choice of the probability distribution of the inter-arrival times should be carefully studied, since it substantially determines the behavior of the system and its performance measurement.
- In queuing systems governed by inter-arrival times with high variability, Hyper-Erlang distribution enables a more accurate modeling than the exponential allows.
- Erlang distribution hyper-demonstrated a significant impact on the modeling of queues, its use is recommended when systems show high variation between arrival times.
- The behavior is the response variables (Lq and Wq) according to the combinations of the factors is the same; although the amounts are different the same patterns and trends is preserved.
- There are changes in the values of Lq and Wq over 100% when making comparisons by level of use among Hyper-Erlang distribution with respect to the Exponential, so your choice becomes a key factor for efficient modeling of a system.
- Changes in factors affecting a queuing system generate a diminishing impact as they increase in their levels. For example, the impact of increasing server has more effect from less servers exist in the system; to change from one to two servers, changing Lq is 45% while passing from 7 to 10, only changes 2% when

inter-arrival times are Hyper-Erlang.

- The factors that most influence the performance of queuing networks are probability distributions between the arrival time, the number of servers in the nodes especially in one- node and system capacity utilization levels near one.
- Systems with Hyper-Erlang distributions have an early destabilization along the nodes even when the utilization level is less than 1.
- For levels below 0.7 use, the ability does not affect system performance.
- Increasing the number of servers with lower use levels of 0.7 has no significant effect on the variation of the measures of system performance.
- The queuing networks in series have bottlenecks at the first node. Thereafter visible decreased queues between nodes whose variation range from 30% to Erlang distribution Híper and 19% for the exponential.
- The results for both distributions differ in the values of performance measures, but overall, the effect of the factors of influence over all combinations shows behaviors and trends are similar for both distributions. This suggests that these behaviors can be all themselves, independent of the type of distribution is used to model inter-arrival times systems.
- Cost minimization is only possible through quantification, due to the existence of intangible costs multi-objective models appear to mediate conflicts between individuals or cost and queue time.
- The game allows optimization through a discount difficulties due to lack of information analysis and implement a system by intelligent travel patterns observed. Include this method as a means of evaluating optimization allows to reach balanced and efficient solutions.
- Optimization models in queuing networks should focus on releasing the first node congestion and ensure a balance in attention span as arrivals at each point of care system.
- Under the same conditions, a country governed by arrivals Hyper-Erlang system needs more resources for optimization than one with exponential times.

- With the development of simulation is possible to know the behavior of queuing networks depending on the combination of the factors that influence it and from that knowledge to feed an optimization model with the aim to efficiently allocate resources.

Analysis of queuing networks may spread to other probability distributions to model the inter-arrival time and attention, since such variations may change the behavior described here.

To find the source of the variation in rates from node to node may be advisable to study dropouts occurring throughout the system and relate to the change in the rate of input nodes within the network.

It is recommended simulation analysis with a probability distribution of the times of Hyper-Erlang attention in order to see the influence this variation in performance measures and conclusions of this study.

For the robustness of the optimization tool it is necessary to study queuing networks under different parameters in order to continue supplying this algorithm with the logic required for operation in a greater number of cases.

BIBLIOGRAPHY

1. Alfaro, Victor M. (2003). Modeling and Analysis of Dynamic Systems Using Generalized Network. In: *Electronics Engineering School of Costa Rica*. p. 25.
2. Anderson David, R. (2004). Quantitative methods for business. *Enterprise Solutions*. Cengage Learning Editores p. 618.
3. Bharucha-Reid, A. T. (1960). Elements of the Theory of Markov Processes and Their appli-cations, McGraw-Hill Book Company, Inc., New York.
4. Burke, P. J. (1976). Proof of a Conjecture on the inter-arrival-Time Distribution in an M/M/1 Queue with Feedback. In: i.e., *EE Transactions on Communications, Vol. 29*. p. 589.
5. Burke, P. J. (2002). The Output of a queuering system, In: Operations Research, Vol 4, p. 704.
6. Cochrana, Jeffery K., & Rocheb, Kevin T. (2009). A multi-class queuing network analysis methodology hospital emergency department for improving performance. In: *Computers & Operations Research*. Vol. 36, p. 138.
7. Cohen, W. A. (2001). How to Make it Big as a Consultant. AMACOM Div American Mgmt Assn.
8. Coss R. Bu. (1982). *Simulation Practical Approach*. Mexico: LIMUSA,

p. 60–65.

9. Dembo, Ron S., Mulvey, John M., & Zenios, Stavros A. (1989). Large-Scale Nonlinear Network Models And Their Application. In *Operations Research. Vol 3*, p. 353.
10. Downb, Bonalda D. (1998). Stability of mixed generalized Jackson networks. In: *School of Industrial and Systems Engineering,* Georgia Institute of Technology, Atlanta, Vol 7, p. 56.
11. Escudero, Laureano F. (1972). Applications of queuing theory: analysis and solution of the problems of bottlenecks in warehouse, production and sales with brief references to Business Games and game theory. First edition. Bilbao: Deusto. p. 259.
12. Evans, Michael J. (2005). Probability and Statistics Editor Reverte. p.675.
13. Feldman, Zohar; Mandelbaum, Avishai; Massey, William A., & Whitt, Ward. (2008). Staffing of Time-Varying Queues to Achieve Time-Stable Performance. In: *Management Science. Vol 54*, p. 2, 3.
14. Francis Comets, Roberto Fernandez, Pablo A. Ferrari. (2002). Processes with Long Memory: Regenerative Construction and Perfect Simulation, Ann. *Appl. Probab. vol. March 12*, 921–943.
15. Gomez, Emilio & Sarabia, José María (2008).. The binomial distribution truncated-exponential with applications in the field of automobile insurance. In: *IAE, Vol 14,* p. 3–20.
16. Gonga, Qiguo & Shouyang, Laib. (2008). Supply chain networks: Closed Jackson network models and properties. In: *Int J. Production Economics, Vol 113*, p. 574.
17. GREEN, Linda and Kolesar, Peter. (1991). The Pointwise Approximation for Queues with Stationary Non stationary Arrivals. In: *Management Science; Vol.37*, p. 84.
18. Gross, Donald & Harris, Carl M. (1998). *Fundamentals of Queuing Theory*. Third edition.
19. H. Berbee. Chains With complete connections: Uniqueness and Markov representation, *Prob Th Rel Fields, 76*, 243–253.
20. Hillier, Frederick S. & Lieberman, Gerald J. (2002). Quantitative methods for management: A modeling approach and case study with spreadsheet. Mexico; Buenos Aires: McGraw-Hill/Irwin, p. 855.
21. Hillier, Frederick S., & Lieberman, Gerald J. (2002). *Operations Research.* Seventh edition. Mexico: McGraw-Hill, p. 1123.

22. Wang, J., Zhou, H., Xu, F., & Li, L. (2005, November). Hyper-Erlang based model for network traffic approximation. In International Symposium on Parallel and Distributed Processing and Applications (pp. 1012-1023). Springer, Berlin, Heidelberg. In: *China Postdoctoral Science Foundation*.

23. Hooker, John Norman. (2007). Integrated methods for optimization. New York: Springer, P. 486.

24. Hyun Jeung, K. O., Chang Seong & Kim Taioun. (2006). A hybrid optimization/simulation approach for a distribution network design of 3PLs. In: *Computers & Industrial Engineering. Vol 50*, p. 586.

25. International Business Report (IBR). Available online: http://www.internationalbusinessreport.com/Reports/Focus-reports/2008_global_overview.asp

26. De Vries, M. G. (1976). *The International Monetary Fund, 1966-1971: Narrative* (Vol. 1). International monetary fund.

27. Jackson, J. R. (1957). Networks of waiting lines. In: *Operation Research. Vol 5*, p. 581.

28. Jackson, J. R. (1963). Queuering Jobshop-like systems. In: *Management Science, Vol 10*, p. 142.

29. Jamhoura, Edgard; Pennaa Manoel; Nabhen, Ricardo & Pujolle, Guy. (2007). Modeling a multi-queue node network with a fuzzy predictor. In: *Fuzzy Sets and Systems*, p. 58.

30. Juran, J. M. (1996). Juran and Quality by Design. p. 75.

31. Kamlesh Mathur & Solow, Daniel. (1996). Operations Research: the art of making decisions. First edition. Mexico; New York: Prentice-Hall Hispanoamericana. p. 977.

32. Kececioglu, Dimitri. (1993). Reliability and Life Testing Handbook. Prentice Hall, Inc. *Volume I*, New Jersey.

33. Kelton, W. David. (2008). *Arena Simulation Software*. 4th ed. Mexico; Bogotá: McGraw-Hill Publishers. p. 631.

34. Kendall, D. G. (1952). Les Processus Stachistiques de croissance en biologie, *Ann. inst. H. Poincare, vol. 13*, pp. 43–108.

35. Kendall. D. G. (1949). Stochastic Processes and Population Growth, J. Roy. *Statist. Soc., ser. B, vol. 11*, pp. 230–264.

36. Kerbache, Lahoucine & MacGregor, James. (2004). Queuing Networks and the topological design of supply chain systems. In: *J. Production*

Economics. Vol 91, p. 278.

37. Kim, Jeong Kim, & Bar. (2007). Regularly varying tails in a queue with autoregressive discrete arrivals of order p. In: *Queuing Syst. Vol 56*, p. 253.
38. Kuehl, Robert. (2000). Statistical analysis for the design and research principles. Thomson Second Edition. p. 20–225.
39. Kurtz, A. D. (1989). *U.S. Patent No. 4,814,845*. Washington, DC: U.S. Patent and Trademark Office.
40. Kuroda, M., & Kawada, A. (1995). Adaptive Control input for job-shop type production Systems with varying Demands using inverse queuing network analysis. In: *Int J. Production Economics. Vol 41*, 356.
41. Lotka, A. J. (1925). *Elements of Physical Biology,* The Williams and Wilkins Company, Baltimore.
42. Malhotra, M., & Reibman, A. (1993). Selecting and implementing phase approximations for semi-Markov models. *Stochastic Models*, *9*(4), 473-506..
43. Mazumdar, R., Mason, L. G., & Douligeris, C. (1991). Fairness in network optimal flow control: Optimality of product forms. *IEEE Transactions on communications*, *39*(5), 775-782.
44. Merchant, D. K., & Nemhauser, G. L. (1978). A model and an algorithm for the dynamic traffic assignment problem. In: *Trans. Sci*, p. 193.
45. Nahmias, Steven. (1999). Analysis of production and operations. First edition. Mexico: CECSA, p. 425. New York; Chichester: John Wiley, (1998). p. 441.
46. Neveu, J. (1977). Processus ponctuels. In *Ecole d'Eté de Probabilités de Saint-Flour VI-1976* (pp. 249-445). Springer, Berlin, Heidelberg.
47. Nogueira Cobra, & Mark Henry. (2000). Marketing services: strategies for tourism, finance, health and communication. Second edition. Bogotá; Buenos Aires: McGraw Hill, p. 255.
48. Nozaki, S., & Ross, S. (1978). Approximations in Finite Capacity Multiserver Queues With Poisson Arrivals, *J. Appl. Prob. 15*.
49. Onicescu, O., & Mihoc, G. (1935). Sur les chaınes statistiques. *CR Acad. Sci. Paris*, *200*, 511-12.
50. Ferrari, P., & Galves, A. (2000). Coupling and regeneration for stochastic processes. In *Notes for a minicourse presented in XIII Escuela Venezolana de Matematicas. Can be downloaded from www.*

ime. usp. br/pablo/book/abstract. html.

51. Philip Kotler, & Gary Armstrong. (2003). *Fundamentals of Marketing*. Pearson Education. p.312.
52. Prawda, Witenberg, Juan. (1976). *Methods and Operations Research Models*. Mexico: Limusa, p. 2V.
53. Render Barry, & Heizer Jay H. (2004). Operations management principles. *Pearson Education*, p. 575
54. Rods, J. S., & Makowski, D. J. K. (1985). queues Competing With linear geometric service requirements and costs: the μc-rule is always optimal. In: *System and letters control. Vol.* 6, p. 207.
55. Ross, S. M. (2014). *Introduction to probability models*. Academic press.
56. Sheldon M. Ross (1983). Stochastics Processes. Wiley Series in Probability and Mathematical Statistics: Probability and Mathematical Statistics. *Lectures in Mathematics, 14*. John Wiley and Sons, Inc., New York.
57. Resnick, S. (1992). Adventures in stochastic processes, Birkhuser Boston. *QA274*.
58. Thompson, G. L., & Zawack, D. J. (1985). A problem expanding parametric programming method for solving the job shop scheduling problem. *Annals of Operations Research*, *4*(1), 327-342.
59. Van Mieghem, J. A. (2000). Price and service discrimination in queuing systems: Incentive compatibility of Gc μ scheduling. *Management Science*, *46*(9), 1249-1267.
60. De Angelis, V., Felici, G., & Impelluso, P. (2003). Integrating simulation and optimisation in health care centre management. *European Journal of Operational Research*, *150*(1), 101-114.
61. Velasco, Enio Edgardo. (1989). Queuing theoretic Approaches for the analysis and synthesis of air traffic control problems. In: *Operation Research. Vol 23*, p. 456.
62. Wang, J., Liu, J., & She, C. (2008). Segment-based adaptive hyper-Erlang model for long-tailed network traffic approximation. The Journal of Supercomputing, 45(3), 296-312.

INDEX